# STUDIES IN ORDINARY DIFFERENTIAL EQUATIONS

*Studies in Mathematics*

The Mathematical Association of America

Volume 1: STUDIES IN MODERN ANALYSIS
*edited by R. C. Buck*

Volume 2: STUDIES IN MODERN ALGEBRA
*edited by A. A. Albert*

Volume 3: STUDIES IN REAL AND COMPLEX ANALYSIS
*edited by I. I. Hirschman, Jr.*

Volume 4: STUDIES IN GLOBAL GEOMETRY AND ANALYSIS
*edited by S. S. Chern*

Volume 5: STUDIES IN MODERN TOPOLOGY
*edited by P. J. Hilton*

Volume 6: STUDIES IN NUMBER THEORY
*edited by W. J. LeVeque*

Volume 7: STUDIES IN APPLIED MATHEMATICS
*edited by A. H. Taub*

Volume 8: STUDIES IN MODEL THEORY
*edited by M. D. Morley*

Volume 9: STUDIES IN ALGEBRAIC LOGIC
*edited by Aubert Daigneault*

Volume 10: STUDIES IN OPTIMIZATION
*edited by G. B. Dantzig and B. C. Eaves*

Volume 11: STUDIES IN GRAPH THEORY, PART I
*edited by D. R. Fulkerson*

Volume 12: STUDIES IN GRAPH THEORY, PART II
*edited by D. R. Fulkerson*

Volume 13: STUDIES IN HARMONIC ANALYSIS
*edited by J. M. Ash*

Volume 14: STUDIES IN ORDINARY DIFFERENTIAL EQUATIONS
*edited by Jack Hale*

***H. A. Antosiewicz***
*University of Southern California*

***Lamberto Cesari***
*University of Michigan*

***Jack Hale***
*Brown University*

***L. K. Jackson***
*University of Nebraska*

***J. P. LaSalle***
*Brown University*

***Jean Mawhin***
*Université Catholique de Louvain*

***M. M. Peixoto***
*Instituto de Matemática Pura e Aplicada, Rio de Janeiro*

***G. R. Sell***
*University of Minnesota*

***Yasutaka Sibuya***
*University of Minnesota*

*Studies in Mathematics*

Volume 14

# STUDIES IN ORDINARY DIFFERENTIAL EQUATIONS

*Jack Hale, editor*

*Brown University*

*Published and distributed by*

The Mathematical Association of America

Library of Congress Catalog Card Number 77-82829

Complete Set ISBN 0-88385-100-8
Vol. 14 ISBN 0-88385-114-8

Printed in the United States of America

Current printing (last digit):

10 9 8 7 6 5 4 3 2 1

# PREFACE

The purpose of this Study in Ordinary Differential Equations is to present some of the areas of current research in such a way as to be accessible to nonspecialists. In addition, each article can serve as a guide to seminars or additional topics for standard courses in differential equations.

Due to the vast nature of the subject, even the selection of topics was very restrictive and we have concentrated on the qualitative theory and the functional analytic aspects of the subject. These important areas, generally, are not easily accessible except through advanced texts and research papers.

The papers of LaSalle, Sell, and Peixoto deal with the general theory of stability and dynamical systems. LaSalle has chosen to present Liapunov functions and invariance in terms of difference equations rather than differential equations. This allows the theory to be presented at a more elementary level without sacrificing any of the essential ideas. The paper of Sell relates the general theory of dynamical systems to nonautonomous differential equations, an important area of current research. The exposition of Peixoto presents generic theory in $R^n$ and, therefore, makes the important theory of structural stability accessible to a much wider audience.

The papers of Jackson, Mawhin, Antosiewicz, and Cesari all deal with the role of functional analysis in the basic theory of differential equations as well as boundary value problems. There is some intentional overlap in the presentations in order to allow each paper to be read independently. Also, it is necessary in the study of differential equations to have many different viewpoints and approaches to similar problems. These four papers should introduce the reader to the methods and problems as well as the manner in which more general abstract theories serve as unification.

The final paper of Sibuya is an excellent introduction to the modern theory of asymptotics. It clearly shows that asymptotics remains a challenging and exciting area of research in differential equations.

JACK K. HALE

# CONTENTS

# STABILITY THEORY FOR DIFFERENCE EQUATIONS*

*J. P. LaSalle*

## 1. INTRODUCTION

Applications in the Soviet Union of Liapunov's direct method to nonlinear problems in the design of control systems in the late 1940's (see [1, 29]) brought about a renewed interest in his theory of stability. The original theory dealt with systems described by ordinary differential equations. An elementary introduction to this theory with some applications is given in [25]. During the past fifteen years there has been a rather rapid development of this theory within, and far beyond, ordinary differential equations. The purpose of this article is to provide an introduction to this theory and to these more recent developments. However, even for ordinary differential equations, this requires an understanding of some of the basic properties of solutions (their existence and

*This article has been influenced by research carried out at the Lefschetz Center for Dynamical Systems during the past decade, and the author wishes to acknowledge the support given to the Center by Brown University, the Air Force Office of Scientific Research, the U. S. Army Research Office, the Office of Naval Research, and the National Science Foundation during this period.

uniqueness, continuous dependence on initial conditions, maximal domains of definition, etc.). None of these difficulties with basic properties are encountered for difference equations, and it turns out to be possible to present through the study of difference equations almost all of the basic features of the general theory, as it stands today, in an elementary and concise manner. This is our objective, and we assume only a good introduction to algebra and analysis in $m$-dimensional Euclidean space $R^m$ (finite dimensional vector spaces and convergence and continuity with respect to Euclidean distance). See, for instance, the Appendix in [12], and [17], Chapters 3 and 5. A good introduction to classical Liapunov theory for difference equations is [22], and [20] goes beyond classical theory and gives some applications to numerical analysis. We go beyond the developments to be found in these two references, and give a number of results that are new for difference equations.

The finite calculus (the calculus of finite differences) and difference equations have been studied as long as the continuous (infinitesimal) calculus and date back to Brook Taylor (1717) and Jacob Stirling (1730). The first treatise was written by L. Euler in 1755. The finite calculus has always had many important applications and was at one time considered to be a prerequisite to infinitesimal calculus. Today, outside of numerical analysis (see [33, 34]) and engineering (system analysis) (see [8, 36]), the finite calculus and difference equations are seldom studied systematically. Elementary textbooks on finite mathematics barely mention the finite calculus and finite difference equations. We learn much more about derivatives, integrals, and Laplace transforms than we do about finite differences, sums, and what engineers call $Z$-transforms. Difference equations are important and significant mathematical models for real phenomena and systems in the physical and nonphysical sciences. After all, the observational data we have for real systems is often discrete, which is why engineers are interested in what they call "sampled-data systems".

In order to make the presentation more complete and more suitable for independent study or an undergraduate seminar, we have included a large number of exercises and references.

## 2. DIFFERENCE EQUATIONS

Let $J$ denote the set of all integers and $J_+$ the set of all nonnegative integers. $R^m$ is real $m$-dimensional Euclidean space; if

$$x = \begin{bmatrix} x_1 \\ x_2 \\ \vdots \\ x_m \end{bmatrix} \in R^m, \ \|x\| = \left(x_1^2 + x_2^2 + \cdots + x_m^2\right)^{1/2},$$

(the Euclidean length of the vector $x$). For a sequence of vectors $x^j$, $x^j \to y$ as $j \to \infty$ means $\|x^j - y\| \to 0$ as $j \to \infty$. We will also use $x$ to denote a function on $J_+$ to $R^m$ ($x : J_+ \to R^m$). Thus we allow the usual and convenient ambiguity where $x$ may denote either a function or a vector (point).

The *first difference* $\dot{x}$ of a function $x$ is defined by

$$\dot{x}(n) = x(n+1) - x(n).$$

Corresponding to the fundamental theorem of calculus, we have $\Sigma_{k=j}^{n} \dot{x}(k) = x(n+1) - x(j)$ and, if $y(n+1) = \Sigma_{k=j}^{n} x(k)$, $\dot{y}(n) = x(n)$. We use $x'$ to denote the function defined by $x'(n) = x(n+1)$ so that $x' = \dot{x} + x$.

A function $T : R^m \to R^m$ is continuous (on $R^m$) if $x^j \to y$ implies $T(x^j) \to T(y)$. We do not need to do so now but later will always assume that $T$ is continuous. If $x$ is a vector, $T(x)$ is a vector and is the value of the function at $x$. If $x$ is a function, the product $Tx$ will denote the composition of functions—$(Tx)(n) = T(x(n))$. Thus the difference equation

$$x(n+1) = T(x(n))$$

can be written

$$x' = Tx. \tag{2.1}$$

*The* solution to the initial value problem

$$x' = Tx, \quad x(0) = x^0, \tag{2.2}$$

is $x(n) = T^n(x^0)$, where $T^n$ is the $n$th-iterate of $T$; $T^0 = I$, the identity function ($Ix = x$), and $T^n = TT^{n-1}$. The solution is defined on $J_+$. There are no difficult questions about the existence and uniqueness of the solutions. Also, it is clear that the solutions are continuous with respect to the initial condition (state) $x^0$ if $T$ is continuous. Unlike ordinary differential equations, the existence and uniqueness is only in the forward direction of time ($n \geqslant 0$). Equation (2.2) is simply an algorithm defining a function $x$ on $J_+$.

Let $g : R^m \to R$. Then

$$u(n+m) = g(u(n), u(n+1), \ldots, u(n+m-1)) \quad (2.3)$$

is an $m$th-order difference equation. The state $u(n+1)$ of the system at time $n+1$ depends on its state at times $n, n-1, \ldots, n-m+1$—i.e., upon this portion of its past history. Note that if we define $x_1(n) = u(n)$, $x_2(n) = x_1'(n) = u(n+1), \ldots, x_m(n) = x_{m-1}'(n) = u(n+m-1)$, then (2.3) is equivalent to the following system of $m$ first-order difference equations (the vector $x(n)$ is the state of (2.3) at time $n+m-1$):

$$\begin{aligned} x_1' &= x_2 \\ x_2' &= x_3 \\ &\vdots \\ x_{m-1}' &= x_m \\ x_m' &= g(x_1, x_2, \ldots, x_m) = g(x) \end{aligned}$$

or

$$x' = Tx,$$

where

$$x = \begin{bmatrix} x_1 \\ x_2 \\ \vdots \\ x_m \end{bmatrix}, \qquad T(x) = \begin{bmatrix} x_2 \\ x_3 \\ \vdots \\ x_m \\ g(x) \end{bmatrix}.$$

For instance, the 3rd-order linear equation

$$u''' + a_2 u'' + a_1 u' + a_0 u = 0$$

is equivalent to

$$x' = Ax,$$

where

$$x_1 = u, \qquad x_2 = u', \qquad x_3 = u'', \qquad x = \begin{bmatrix} x_1 \\ x_2 \\ x_3 \end{bmatrix},$$

and

$$A = \begin{bmatrix} 0 & 1 & 0 \\ 0 & 0 & 1 \\ -a_0 & -a_1 & -a_2 \end{bmatrix},$$

a $3 \times 3$ matrix. The initial conditions $u(0) = u_0$, $u'(0) = u_0'$, $u''(0) = u_0''$ correspond to

$$x(0) = \begin{bmatrix} u_0 \\ u_0' \\ u_0'' \end{bmatrix} = x^0.$$

The solution of this initial value problem is

$$x(n) = A^n x^0.$$

We are often interested in obtaining information about the asymptotic behavior of solutions; i.e., what happens to solutions for large $n$? For instance, in the above example, when does $A^n \to 0$ as $n \to \infty$? ($A^n = (a_{ij}^{(n)})$ and $A^n \to 0$ means $a_{ij}^{(n)} \to 0$ for all $i, j$ or, equivalently, $A^n x \to 0$ for each $x$.)

## 3. POSITIVE LIMIT SETS OF SOLUTIONS AND INVARIANCE

We now adopt the following notations: For $S$ a subset of $R^m$, $\bar{S}$ is the closure of $M$. If $x \in R^m$ and $S$ is a closed set of $R^m$, then $\rho(x, S) = \min\{\|x - y\|; y \in S\}$, the *distance of $x$ from $S$*. $T^n(x^0) \to S$ as $n \to \infty$ means that $\rho(T^n(x^0), S) \to 0$ as $n \to \infty$.

Thus, if we can locate a closed set that a solution $T^n(x^0)$ approaches as $n \to \infty$, we have obtained asymptotic information about the solution. We shall show that Birkhoff's positive limit set ($\omega$-limit set) of a solution (introduced in [6]) is the smallest closed set that $T^n(x^0)$ approaches as $n \to \infty$. In this sense, locating a solution's positive limit set is the "best" asymptotic information we can hope to obtain. Our objective is to learn that the role of Liapunov functions is to help locate positive limit sets of solutions. To that end, we proceed to define limit sets and to study their basic properties.

3.1. DEFINITION (Birkhoff): A point $y$ is a positive limit point of $T^n x$ if there is a sequence of integers $n_i$ such that $n_i \to \infty$ and $T^{n_i}x \to y$ as $i \to \infty$. The *positive limit set* $\Omega(x)$ of $T^n x$ is the set of all its positive limit points.

3.2. *Exercise.* Show that an alternate definition for $\Omega(x)$ is

$$\Omega(x) = \bigcap_{j=0}^{\infty} \overline{\bigcup_{j=n}^{\infty} T^n x}.$$

3.3. *Exercise.* Let $S_n$ be a sequence of closed (compact) nonincreasing ($S_{n+1} \subset S_n$) sets in $R^m$. Show that $\cap_{j=0}^{\infty} S_n$ is closed (nonempty and compact).

3.4. DEFINITION: Relative to (2.1), or to $T$, a set $H \subset R^m$ is said to be *positively* (*negatively*) *invariant* if $T(H) \subset H$ ($H \subset T(H)$). $H$ is said to be *invariant* if $T(H) = H$; i.e., it is both positively and negatively invariant.

Now to have an interesting theory we must put a smoothness condition on $T$. All that we need is to assume that $T$ is continuous, and we do so *from this point on*. As a model for the real world, at least, we expect continuity.

For ordinary differential equations whose solutions are continuous curves, one is able to conclude that positive limit sets of bounded solutions are connected. Here we cannot expect this. The concept of connectedness has to be related to invariance.

3.5. DEFINITION: A closed invariant set $H$ is said to be *invariantly connected* if it is not the union of two nonempty disjoint invariant closed sets.

3.6. DEFINITION: A solution $T^n x$ is said to be *periodic* (or *cyclic*) if for some $k > 0$, $T^k x = x$. The least such $k$ is called the *period* of the solution or the order of the cycle. If $k = 1$, $x$ is a *fixed point* of $T$ or an *equilibrium state* of (2.1).

Note here that, unlike ordinary differential equations, a solution can reach an equilibrium state in finite time.

Periodic phenomena (oscillations) are of theoretical and practical interest, and the next exercise shows a relationship between periodicity and invariant connectedness.

3.7. *Exercise.* Show that: A finite set (a finite number of points) is invariantly connected if and only if it is a *periodic motion*. (*Hint*: Any permutation can be written as a product of disjoint cycles. See, for instance, [7], p. 140, Theorem 11.)

3.8. *Exercise.* Show that:

(a) The closure of a positively invariant set is positively invariant.

(b) The closure of a bounded invariant set is invariant.

As the above exercise suggests, the closure of an unbounded invariant set may not be invariant.

3.9. DEFINITION: $T_n x$ (defined for all $n \in J$) is called an *extension of the solution* $T^n x$ *to* $J$ if $T_0 x = x$ and $T(T_n x) = T_{n+1} x$ for all $n \in J$.

Note that $T_n x = T^n x$ for all $n \in J_+$, and that, if $x$ is in an invariant set $H$, then $T^n x$ has an extension on $J$, which may or may not be unique.

3.10. *Exercise.* Show that a set $H$ is invariant if and only if each motion starting in $H$ has an extension on $J$ that is in $H$ for all $n$.

An invariant set $H$ can have an extension on $J$ from a point in $H$ that is not in $H$. However, the above exercise says it always has at least one that does not leave $H$.

3.11. *Exercise.* Let $E$ be a set in $R^m$, and let $M$ be the largest (by inclusion) invariant set in $E$. Show that:

(a) $M$ is the union of all extended solutions that remain in $E$ for all $n$.

(b) $x \in M$ if and only if there is an extended solution from $M$ in $E$ for all $n$.

(c) If $E$ is compact (bounded and closed), $M$ is compact.

At last we are in a position to establish the basic properties of positive limit sets. The most interesting case is when $T^n(x)$ is bounded for all $n \geqslant 0$. Such motions are often said to be *positively stable in the sense of Lagrange.*

3.12. THEOREM: *Every positive limit set is closed and positively invariant. If $T^n(x)$ is bounded for all $n \in J_+$, then $\Omega(x)$ is nonempty, compact, invariant, invariantly connected, and is the smallest closed set that $T^n x$ approaches as $n \to \infty$.*

*Proof:* It is easily seen that the complement of $\Omega(x)$ is open, and hence $\Omega(x)$ is closed (see also Exercises 3.2 and 3.3). If $q \in \Omega(x)$, then there is a sequence $n_i$ such that $n_i \to \infty$ and $T^{n_i}x \to q$ as $i \to \infty$. By continuity of $T$, $T(T^{n_i}(x)) = T^{n_i+1}(x) \to T(q)$, and $T(q) \in \Omega(x)$. Hence $T(\Omega(x)) \subset \Omega(x)$, and $\Omega(x)$ is positively invariant.

If $T^n(x)$ is bounded, it contains a convergent subsequence and $\Omega(x)$ is nonempty. Also $\Omega(x) \subset \overline{\cup_{n=0}^{\infty} T^n(x)}$, and is, therefore, bounded and hence compact. If $q \in \Omega(x)$, we now have the existence of a sequence $n_i$ such that $n_i \to \infty$, $T^{n_i}x \to q$ and $T^{n_i-1}x \to p$ as $i \to \infty$. Then $T(T^{n_i-1}x) = T^{n_i}(x) \to T(p) = q$. Therefore, $\Omega(x) \subset T(\Omega(x))$, and $\Omega(x)$ is invariant when $T^n(x)$ is bounded.

We will show next that $T^n(x) \to \Omega(x)$ as $n \to \infty$ if $T^n(x)$ is bounded. Clearly $\rho(T^n(x), \Omega(x))$ is bounded. If $\rho(T^n(x), \Omega(x))$ does not approach 0 as $n \to \infty$, there is a subsequence $n_i$ such that $n_i \to \infty$, $T^{n_i}(x)$ converges and $\rho(T^{n_i}(x), \Omega(x))$ does not converge to

0. Clearly this leads to a contradiction, since the limit of $T^{n_i}(x)$ is in $\Omega(x)$, and $T^n(x) \to \Omega(x)$ as $n \to \infty$. If $T^n(x) \to E$ and $E$ is closed, then it is easy to see that $\Omega(x) \subset E$. Hence $\Omega(x)$ is the smallest closed set that $T^n(x)$ approaches.

It remains to show that if $T^n(x)$ is bounded, then $\Omega(x)$ is invariantly connected. Assume $\Omega(x)$ is the union of two disjoint closed nonempty invariant sets $\Omega_1$ and $\Omega_2$. Since $\Omega(x)$ is compact, so are $\Omega_1$ and $\Omega_2$. There then exist disjoint open sets $U_1$ and $U_2$ such that $\Omega_1 \subset U_1$ and $\Omega_2 \subset U_2$. Also, since $T$ is uniformly continuous on $\Omega_1$, there is an open set $V_1$ such that $\Omega_1 \subset V_1$ and $T(V_1) \subset U_1$. Since $\Omega(x)$ is the smallest closed set that $T^n(x)$ approaches $T^n(x)$ must intersect both $V_1$ and $U_2$ an infinite number of times. But this implies the existence of a convergent subsequence $T^{n_i}x$ that is not in either $V_1$ or $U_2$. This contradiction, since $T^{n_i}(x)$ has a convergent subsequence, shows that $\Omega(x)$ is invariantly connected and completes the proof.

3.13. *Exercise.* Show that: If $K$ is compact and positively invariant, then $\cap_{n=0}^{\infty} T^n(K)$ is nonempty, compact and invariant, and is obviously the largest invariant set in $K$.
(*Suggestion*: This can be proved, and more is learned, by first studying the properties of

$$\Omega(H) = \bigcap_{j=0}^{\infty} \overline{\bigcup_{n=j}^{\infty} T^n(H)}\,.$$

Note that $y \in \Omega(H)$ if and only if there are sequences $n_j \in J_+$ and $y^j \in H$ such that $T^{n_j}y^j \to y$ and $n_j \to \infty$ as $j \to \infty$.)

## 4. LIPAUNOV FUNCTIONS AND AN EXTENSION OF LIAPUNOV'S DIRECT METHODS

What we will do here is to so define the concept of a Liapunov function that we obtain, exploiting the invariance property of positive limit sets, a result which relates Liapunov functions to the location of positive limit sets. Because of this, the principal result is called an "invariance principle". Our definition is much less restrictive than Liapunov's and greatly extends his direct method.

(His method is called "direct" because it does not depend on a knowledge of solutions.) This was first done for autonomous ordinary differential equations in [24] (see also [26]), and today has been extended to infinite dimensional dynamical systems (difference-differential equations, functional differential equations, certain types of partial differential equations and evolutionary equations (for example, see [16]). It also has been extended to nonautonomous systems (references are given later). It is an elementary result but one with many important applications.

Let $V : R^m \to R$. Relative to (2.1) (or to $T$) define

$$\dot{V}(x) = V(T(x)) - V(x).$$

If $x(n)$ is a solution of (2.1),

$$\dot{V}(x(n)) = V(x(n+1)) - V(x(n));$$

$\dot{V}(x) \leqslant 0$ means that $V$ is nonincreasing along solutions.

4.1. Definition: Let $G$ be any set in $R^m$. We say that $V$ is a *Liapunov function of* (2.1) *on* $G$ if (i) $V$ is continuous on $R^m$, and (ii) $\dot{V}(x) \leqslant 0$ for all $x \in G$.

For $V$ a Liapunov function of (2.1) on $G$, we define

$$E = \left\{x;\ \dot{V}(x) = 0,\ x \in \overline{G}\right\}$$

($\overline{G}$ is the closure of $G$). We use $M$ to denote the largest invariant set in $E$ (see Exercise 3.11). $V^{-1}(c) = \{x;\ V(x) = c,\ x \in R^m\}$, a level surface.

4.2. Theorem (Invariance Principle): *If* (i) *$V$ is a Liapunov function of* (2.1) *on $G$, and* (ii) *$x(n)$ is a solution of* (2.1) *bounded and in $G$ for all $n \geqslant 0$, then there is a number $c$ such that*

$$x(n) \to M \cap V^{-1}(c).$$

*Proof:* The proof now is quite easy. By our assumptions $V(x(n))$ is nondecreasing with $n$ and is bounded from below.

Hence, $V(x(n)) \to c$ as $n \to \infty$. If $p \in \Omega(x(0))$, then there is a sequence $n_i$ such that $n_i \to \infty$ and $x(n_i) \to p$. By continuity of $V$, $V(x(n_i)) \to V(p) = c$ and $\Omega(x(0)) \subset V^{-1}(c)$. Since $\Omega(x(0))$ is invariant, $V(T^n(p)) = c$ and $\dot{V}(p) = 0$. Therefore, $\Omega(x(0)) \subset E$, and hence in $M$. This completes the proof.

The difficulty in applications is to find "good" Liapunov functions—ones that make $M$ as small as possible. For instance, a constant function is always a Liapunov function but gives no information.

Let us look at a simple example which illustrates how the result is applied. Consider the 2-dimensional system

$$x(n+1) = \frac{ay(n)}{1 + x^2(n)},$$

$$y(n+1) = \frac{bx(n)}{1 + y^2(n)},$$

or

$$x' = \frac{ay}{1 + x^2}, \qquad y' = \frac{bx}{1 + y^2}. \tag{4.1}$$

Take $V(x, y) = x^2 + y^2$. Then

$$\dot{V}(x, y) = \left( \frac{b^2}{(1 + y^2)^2} - 1 \right) x^2 + \left( \frac{a^2}{(1 + x^2)^2} - 1 \right) y^2.$$

Case 1. $a^2 < 1, b^2 < 1$. Now

$$\dot{V} \leqslant (b^2 - 1)x^2 + (a^2 - 1)y^2,$$

and $V$ is a Liapunov function of (4.1) on $R^2$. Here $M = E = \{(0, 0)\}$, and since every solution is clearly bounded, we have by

Theorem 4.2 that every solution approaches the origin as $n \to \infty$ (the origin is a global attractor and later, as we shall explain, we can conclude in this case that the origin is globally asymptotically stable). This is Liapunov's classical case—$V(x)$ and $-\dot{V}(x)$ are positive definite.

CASE 2. $a^2 \leqslant 1$, $b^2 \leqslant 1$ and $a^2 + b^2 < 2$. We may assume that $a^2 < 1$ and $b^2 = 1$. $V$ is still a Liapunov function of (4.2) on $R^2$ but $-\dot{V}$ is not positive definite. In fact, $\dot{V} \leqslant (a^2 - 1)y^2$, and $E$ is the $x$-axis ($y = 0$). However, since $T(x, 0) = (0, bx)$, we see that again $M$ is just the origin. The conclusion is the same as in Case 1.

CASE 3. $a^2 = b^2 = 1$. $V$ is still a Liapunov function of (4.2) and all solutions are still bounded. Here $E = M$ is the union of the two coordinate axes, and by Theorem 4.2 we know that each solution approaches $\{(c, 0), (0, c), (-c, 0), (0, -c)\}$—the intersection of $E$ with the circle $x^2 + y^2 = c^2$. There are two subcases.

(i) $ab = 1$. Then $T(c, 0) = (0, bc)$, $T^2(c, 0) = T(abc, 0) = (c, 0)$. Since positive limit sets are invariantly connected, every solution approaches one of these periodic motions—the origin or a periodic motion of period 2 (see Exercise 3.7).

(ii) $ab = -1$. Here $T(c, 0) = (0, bc)$, $T^2(c, 0) = (abc, 0) = (-c, 0)$, $T^3(c, 0) = (0, -bc)$, and $T^4(c, 0) = (-abc, 0) = (c, 0)$. If $c \neq 0$, these are periodic motions of period 4. As in (i) above, each solution approaches the origin or one of these periodic motions of period 4.

CASE 4. $a^2 > 1$, $b^2 > 1$. Let $B_\delta = \{(x, y);\ x^2 + y^2 < \delta^2\}$. For $x \in B_\delta$ and $\delta$ sufficiently small

$$\dot{V} \geqslant \left(\frac{b^2}{1 + \delta^2} - 1\right)x^2 + \left(\frac{a^2}{1 + \delta^2} - 1\right)y^2 > 0,$$

and $-V$ is a Liapunov function of (4.2) on $B_\delta$ for $\delta$ sufficiently small, and $E = M = \{(0, 0)\}$. No solution starting at a point in $B_\delta$ other than the origin can approach the origin from within $B_\delta$ (its distance from the origin is increasing) and $T(x, y) = (0, 0)$ implies $x = y = 0$. Therefore each such solution must leave $B_\delta$ by Theorem 4.2 (instability) and, since no solution can jump to the origin in finite time except the trivial solution, there is no nontrivial solution that can approach the origin as $n \to \infty$. This is called strong instability (see Section 5). There is not much more information that can be obtained from this particular Liapunov function. The only cases that could be handled by Liapunov's classical direct method are $a^2 < 1$ and $b^2 < 1$, and $a^2 > 1$ and $b^2 > 1$.

## 5. STABILITY AND INSTABILITY

The original definition by Liapunov (see [30]) for stability was for solutions (motions) and can be viewed as an asymptotic continuity with respect to initial conditions. We will restrict ourselves to stability of equilibrium states or positively invariant sets (equilibrium sets).

5.1. DEFINITION: A set $H$ is said to be *stable* if given a neighborhood $U$ of $H$ (an open set containing $\bar{H}$), there is a neighborhood $W$ of $H$ such that $T^n(W) \subset U$ for all $n \in J_+$ $(T(W) = \{T(x); x \in W\})$.

If $H$ is bounded, a complete neighborhood system of $H$ is given by the spheres $B_\varepsilon(H) = \{x; \rho(x, H) < \varepsilon\}$.

5.2. *Exercise.* Show that: If $H$ is stable, then $\bar{H}$ is positively invariant. In particular, if $H$ is a point, it is an equilibrium point.

For $H$ a set in $R^m$ define $\hat{H}$ as follows: $z \in \hat{H}$ if there is a sequence $x_i \in R^m$ and a sequence $n_i \in J_+$ such that $x_i \to y \in \bar{H}$ and $T^{n_i}(x_i) \to z$. In topological dynamics, the set of all such $z$ for which $n_i$ is bounded is called the *prolongation* of $H$ (if $H$ is a point $x$, $\hat{H}$ is called a prolongation of the solution $T^n x$). The set of all such $z$ for which $n_i \to \infty$ is called the *prolongation limit set* of $H$. Note that $H \subset \hat{H}$.

5.3. LEMMA: *Let $H$ be a compact positively invariant set. Then*
(a) *$\hat{H}$ is invariant,*
(b) *$H$ is stable if and only if $\hat{H} = H$.*

*Proof:* The proof of (a) is about the same as the proof of invariance in Theorem 3.12.

It is also clear that, if there is a $z \in \hat{H}$ that is not in $H$, then $H$ is not stable. Suppose $H$ is not stable. Then for some bounded neighborhood $U$ of $H$ there exists a sequence $x^i$ such that $x^i \to y \in H$ as $i \to \infty$ and each motion $T^n x^i$ eventually leaves $U$. Let $n_i$ be the smallest integer with the property that $T^{n_i} x^i$ is not in $U$. Since $T$ is continuous, $T(U)$ is bounded and so is the sequence $T^{n_i} x^i$. It, therefore, has a convergent subsequence which converges to a point not in $U$, and hence $H$ not stable implies $\hat{H} \neq H$. This completes the proof.

The stability concept of greatest practical interest is "asymptotic stability". We will see why in Section 9. We now define this type of stability.

5.4. DEFINITION: A set $H$ is an *attractor* if there is a neighborhood $U$ of $\bar{H}$ such that $x \in U$ implies $T^n x \to \bar{H}$ as $n \to \infty$. It is a *global attractor* if $T^n x \to H$ for all $x \in R^m$. If $H$ is both stable and an attractor, then $H$ is said to be *asymptotically stable*. If $H$ is stable and is a global attractor, $H$ is said to be *globally asymptotically stable*. *Unstable* means not stable. If $H$ is neither stable nor an attractor, we will say that $H$ is *strongly unstable*.

5.5. *Exercise.* Given a set $H$, its inverse image $T^{-1}(H) = \{y; T(y) \in H\}$. A set is said to be *inversely invariant* if $T^{-1}(H) = H$. Show that:

(a) A set $H$ is inversely invariant if and only if $H$ is positively invariant and $T^{-1}(H) \subset H$.

(b) A set $H$ is negatively invariant if and only if $T^{-1}(x)$ intersects $H$ for each $x \in H$.

For ordinary differential equations inverse invariance and invariance are identical (there is uniqueness of solutions in both directions of time).

5.6. *Exercise. The region of attraction* $\mathcal{R}(H)$ of a set $H$ is the set of all $x$ such that $T^n x \to H$ as $n \to \infty$. The boundary of $H$ is denoted by $\partial H$ and its complement by $\mathcal{C}H$ $(\partial H = \bar{H} \cap \overline{\mathcal{C}H})$.

Show that: If $H$ is asymptotically stable, then

(a) $\mathcal{R}(H)$ is open,

(b) $\mathcal{R}(H)$, $\partial H$, and $\mathcal{C}\mathcal{R}(H)$ are inversely invariant.

We now have as an almost immediate consequence of Theorem 4.2 and Lemma 5.3:

5.7. THEOREM: *Let $G$ be a bounded open positively invariant set. If* (i) *$V$ is a Liapunov function of* (2.1) *on $G$,* (ii) *$M \subset G$, then $M$ is an attractor and $G \subset \mathcal{R}(M)$. If, in addition,* (iii) *$V$ is constant on $M$, then $M$ is asymptotically stable (globally asymptotically stable relative to $G$).*

*Proof:* The first part of the theorem is an immediate consequence of Theorem 4.2. $E$ is closed and in $\bar{G}$ and hence is compact. Therefore, $M$ is compact (Exercise 3.11). Assume now that (iii) holds. Then by Lemma 5.3 $M$ not stable implies $\hat{M} \neq M$. However, it is not difficult to see that $V(x) = c$ for $x \in \hat{M}$ and since $\hat{M}$ is invariant, $\hat{M} \subset E$. This contradicts the definition of $M$, which is the largest invariant set in $E$ and completes the proof.

Note that, if $M$ is a single point, then (iii) is automatically satisfied. For example, we can now conclude global asymptotic stability of (4.1) if $a^2 \leqslant 1$, $b^2 \leqslant 1$, and $a^2 + b^2 < 2$. This is also true if $M$ is a finite invariantly connected set (see Exercise 3.7). Note also that if no solution outside $M$ in $G$ can reach $M$ in finite time (i.e., $M$ is also inversely invariant), then condition (iii) can be replaced by: (iii)′ $V$ is constant on the boundary of $M$. Another point of interest in applications is that the size of $G$ gives information as to the "extent" of asymptotic stability.

5.8. *Exercise.* Show that: If (i), (ii), and (iii) above are satisfied and $M$ is the largest positively invariant set in $E$, then $V(x) > c$ for each $x \in G - M$, where $c$ is the value of $V$ on $M$ (i.e., $V(x) - c$ is positive definite relative to $M$).

It turns out in applications that the largest positively invariant set in $E$ is usually the same as the largest invariant set in $E$ (examples to the contrary are quite artificial) and "good" Liapunov functions will usually be positive definite. However, except for quadratic forms, where there are computable criteria (see [21]), positive definiteness may be difficult to establish, even when $M$ is a point. Theorem 5.7 tells us that it is not necessary to verify that $V$ is positive definite with respect to $M$. Rather, Theorem 5.7 plus Exercise 5.8 gives a sufficient condition for positive definiteness ($c = 0$) (when $M$ is a single point, this is a sufficient condition for the existence of a local minimum of $V$).

5.9. *Exercise* (The analog of Liapunov's first two theorems on stability). Let 0 be an equilibrium point of (2.1), and let $N$ be a neighborhood of the origin. $V : R^n \to R$ is said to be positive definite if $V(x) \geqslant 0$ on $N$ and $V(x) = 0$ and $x \in N$ implies $x = 0$ (the origin is an isolated minimum of $V$). Show that: If (i) $V$ positive definite and (ii) $\dot{V}$ is a Liapunov function of (2.1) on $N$, then the origin is stable. If, in addition, $\dot{V}$ is negative definite (i.e., $-V$ is positive definite), the origin is asymptotically stable. *Suggestion*. You may assume $N$ is bounded. Let $2m_0$ be the minimum of $V$ on the boundary of $N$. Let $G = \{x;\ V(x) < m_0;\ x \in N\}$. Argue that $G$ is positively invariant and apply Theorem 5.7.

5.10. *Exercise*. Generalize the result in Exercise 5.9 for the stability of compact positively invariant sets.

We could develop specialized criteria for instability but prefer to place emphasis on the use of Theorem 4.2. This was illustrated in our discussion of (4.1). We look at another example which is a bit more interesting. The technique we will use generalizes that given by Ĉetaev for ordinary differential equations. He wanted to show for conservative dynamical systems that, if an equilibrium is not a minimum of the potential energy, then it is not stable (see [25], p. 56). Lagrange had enunciated the converse—at a minimum the equilibrium is stable—and Liapunov proved it. A Liapunov function is a generalized energy function. Consider

$$\begin{aligned} x' &= f(x, y), \\ y' &= g(x, y), \quad f(0, 0) = g(0, 0) = 0. \end{aligned} \tag{5.1}$$

Let $V = -xy$. Then $\dot{V} = xy - f(x, y)g(x, y)$. Let $G = \{(x, y);$

$xy > 0, x^2 + y^2 < \delta^2\}$. Assume that $f(x, y)g(x, y) > 0$ for $xy > 0$ and that $\dot{V} < 0$ for $(x, y) \in G$. Note that $xy > 0$ is positively invariant and that $V$ is a Liapunov function on $G$. Since $V$ vanishes on $x = 0, y = 0$, and no solution starting in $G$ can remain in $G$ and approach these axes ($\dot{V} < 0$ in $G$), no solution starting in $G$ can remain in $G$ for all $n > 0$ (this would contradict Theorem 4.2). Also when a solution leaves $G$ it goes outside $x^2 + y^2 < \delta^2$. Hence the origin is not stable. Since $xy > 0$ is positively invariant, the origin is not an attractor, and it is strongly unstable—$\delta$ is a measure of its instability.

## 6. LIAPUNOV FUNCTIONS

Everyone who has worked with Liapunov functions knows that two Liapunov functions are better than one, and except for notational convenience there is very little gained by the usual concept of a vector Liapunov function. The term vector Liapunov function seems to have been first used by Bellman [5]. For an application where again the terminology is used and other references see [14]. We will generalize the usual concept of a vector Liapunov function. The idea comes from a construction of a Liapunov function by the economist Arrow *et al.* in [2], and this has been much exploited by economists in studying stability in competitive analysis. They, however, seem not yet to be acquainted with the extensions of Liapunov's direct method and the use of an invariance principle. For applications of stability theory in economics see [2, 3, 18, 19]. An interesting difference equation model for the control of unemployment and inflation is given in [40].

For $x \in R^m$, $x > 0$ means $x_i > 0$ and $x \geqslant 0$ means $x_i \geqslant 0$, $i = 1, \ldots, m$. Let $v : R^m \to R^q$ and define $\dot{v}(x) = v(T(x)) - v(x)$. All definitions and result translate exactly, and we have the exact analog of Theorem 4.2. Now, of course, if $v$ is a vector Liapunov function—in this the usual sense—then each $v_i$ is a scalar Liapunov function for each $i$ and so is $V = \Sigma_{i=1}^{q} v_i$. The set $M$ for $v$ is the same as the set $M$ for $V$, but there may be a difference between $M \cap v^{-1}(c)$ and $M \cap V^{-1}(c)$. So there may be some information gained by the use of the vector Liapunov $v$ function but this has not been demonstrated by a significant example. In any case what we want to do is to go to something more general.

For a vector function $w : R^m \to R^q$, define $W(x) = \max_i w_i(x)$. If $w$ is continuous, then so is $W$. Define $\mathring{w}(x) = w(T(x)) - W(x)u$, where $u_i = 1$ for $i = 1, 2, \ldots, q(\mathring{w}_i(x) = w_i(T(x)) - W(x))$.

6.1. DEFINITION: We will say that $w$ is a *vector Liapunov function* of $x' = T(x)$ on $G$ if (i) $w$ is continuous and (ii) $\mathring{w}(x) \leqslant 0$ for all $x \in G$.

We define $E = \{x; \mathring{w}(x) \nless 0, x \in \overline{G}\}$; i.e., $x \in E$ if $x \in \overline{G}$ and $\mathring{w}_i(x) = 0$ for some $i = 1, \ldots, q$. $M$ is the largest invariant set in $E$. Note that

$$\mathring{w}(x) = \dot{w}(x) + w(x) - W(x)u$$

so that $\mathring{w}(x) \leqslant \dot{w}(x)$ for all $x$, and requiring $\mathring{w} \leqslant 0$ is not as strong as $\dot{w} \leqslant 0$. In fact, $w$ can be a vector Liapunov function in our sense, and yet it may be that no component of $w$ is a scalar Liapunov function. However, $\dot{W}(x) = \operatorname{Max}_i \mathring{w}_i(x)$ so that, if $w$ is a vector Liapunov function, then $W$ is a scalar Liapunov function. This is just another, and a good way, for constructing a scalar Liapunov function from a number of scalar functions. This idea would seem to have natural applications to problems in control where you wish always to be sure that at each time you reduce the largest component of a vector measure $w$ of the error in control (or performance). The use of a vector Liapunov function of this type to design a control that does this may be an idea that is worth exploring. We have immediately from Theorem 4.2

6.2. COROLLARY: *If* (i) *$w$ is a vector Liapunov function of* (2.1) *on $G$ and* (ii) *a solution $x(n)$ of* (2.1) *is in $G$ and bounded for all $n \geqslant 0$, then, for some $c$, $x(n) \to M \cap W^{-1}(c)$ as $n \to \infty$.*

We illustrate the use of this result in the next section. For an application to the study of an epidemic model see [28].

## 7. LINEAR DIFFERENCE EQUATIONS

Let $A = (a_{ij})$ be a real $m \times m$ matrix. $A^T = (a_{ji})$, the transpose of $A$. $A^n = AA^{n-1}$, $A^0 = I$. If $\lambda_1, \lambda_2, \ldots, \lambda_m$ are the eigenvalues of $A$, $\sigma(A) = \{\lambda_1, \lambda_2, \ldots, \lambda_m\}$ (the *spectrum* of $A$) and $r(A) = \text{Max}_i|\lambda_i|$ (the *spectral radius*).

The general linear difference equation of dimension $m$ is

$$x' = Ax. \tag{7.1}$$

The solution satisfying $x(0) = x^0$ is $A^n x^0$. The columns of $A$ are the *principal solutions* of (7.1). If $v^i$ is an eigenvector of $A$ with eigenvalue $\lambda_i$, then $c_i\lambda_i^n v^i$ is a solution of (7.1). If the eigenvalues of $A$ are distinct, then the general solution of (7.1) is

$$x(n) = c_1\lambda_1^n v^1 + \cdots + c_n\lambda_q^n v^q.$$

Thus, if $r(A) \geqslant 1$, there is always a solution that does not approach the origin. If $r(A) > 1$, there are unbounded solutions.

**7.1 An algorithm for computing $A^n$ from its eigenvalues.** This algorithm is the analog of Putzer's algorithm in [35] for computing $e^{At}$.

We look for a representation of $A^n$ in the form

$$A^n = \sum_{j=1}^{m} w_j(n)Q_{j-1} \tag{7.2}$$

where

$$Q_j = (A - \lambda_j I)Q_{j-1},\ Q_0 = I. \tag{7.3}$$

$Q_m = 0$ by the Hamilton-Cayley Theorem (every matrix satisfies its characteristic equation). It is just this fact that suggests the form of the representation (7.2).

The initial condition $A^0 = I$ is satisfied by taking $w_1(0) = 1$, $w_2(0) = \cdots = w_m(0) = 0$. We want $A\Sigma_{j=1}^m w_n(n)Q_{j-1} = \Sigma_{j=1}^m w_j(n+1)Q_{j-1}$ or, since $AQ_{j-1} = Q_j + \lambda_j Q_{j-1}$,

$$\sum_{j=1}^{m} w_j(n)(Q_j + \lambda Q_{j-1}) = \sum_{j=1}^{m} w_j(n+1)Q_{j-1}.$$

Thus, (7.2) holds if

$$\begin{aligned} &w_1(n+1) = \lambda_1 w_1(n),\ w_1(0) = 1,\ (w_1(n) = \lambda_1^n) \\ &w_j(n+1) = \lambda_j w_j(n) + w_{j-1}(n),\ w_j(0) = 0, j = 2, \dots, m. \end{aligned} \tag{7.4}$$

Equations (7.3) and (7.4) are algorithms for computing the $Q_j$ and the $w_j(n)$ in terms of the eigenvalues of $A$ (i.e., for computing $A^n$ if we know or have computed the eigenvalues of $A$).

By way of illustration let us use the algorithm to find the solution of

$$y''' - 3y'' + 3y' - y = 0; \quad y''(0) = y_0'', y'(0) = y_0', y(0) = y_0.$$

This 3rd-order equation is equivalent to $x' = Ax$ where

$$x = \begin{bmatrix} x_1 \\ x_2 \\ x_3 \end{bmatrix} = \begin{bmatrix} y \\ y' \\ y'' \end{bmatrix} \quad \text{and} \quad A = \begin{bmatrix} 0 & 1 & 0 \\ 0 & 0 & 1 \\ 1 & -3 & 3 \end{bmatrix}.$$

$$\phi(\lambda) = \det(A - \lambda I) = -(\lambda - 1)^3 \quad \text{and} \quad \lambda_1 = \lambda_2 = \lambda_3 = 1.$$

$$Q_0 = I,\ Q_1 = A - I = \begin{bmatrix} -1 & 1 & 0 \\ 0 & -1 & 0 \\ 1 & -3 & 2 \end{bmatrix},$$

$$Q_2 = (A - I)^2 = \begin{bmatrix} 1 & -2 & 1 \\ 1 & -2 & 1 \\ 1 & -2 & 1 \end{bmatrix}.$$

Solving (7.4) directly, or by using Exercise 7.3, we obtain $w_1(n) = 1$, $w_2(n) = n$, $w_3(x) = \frac{1}{2}n(n-1)$. Hence

$$A^n = I + n(A - I) + \tfrac{1}{2}n(n-1)(A - I)^2$$

$$= \begin{bmatrix} \frac{1}{2}(n-1)(n-2) & -n(n-2) & \frac{1}{2}n(n-1) \\ \frac{1}{2}n(n-1) & -(n+1)(n-1) & \frac{1}{2}(n+1)n \\ \frac{1}{2}(n+1)n & -(n+2)n & \frac{1}{2}(n+2)(n+1) \end{bmatrix}.$$

The solution $y(n)$ is the first component of $A^n x^0$. This gives $y(n) = \frac{1}{2}(n-1)(n-2)y_0 - n(n-2)y_0' + \frac{1}{2}n(n-1)y_0''$.

7.2. *Exercise.* (a) Show that: If $r(A) = r_0 < \beta$ then

$$|w_j(n)| \leqslant \frac{\beta^n}{(\beta - r_0)^{j-1}}, \qquad j = 1, \ldots, m.$$

(b) if $\alpha < \min_i |\lambda_i|$, establish a lower bound for $|w_j(n)|$.

From Exercise 7.2 we see that, if $|r(A)| < 1$, then $w_j(n) \to 0$ as $n \to \infty$, and hence $A^n \to 0$ as $n \to \infty$. We already know when $r(A) \geqslant 1$ that $A^n$ does not approach 0 as $n \to \infty$. $A^n \to 0$ corresponds to the global asymptotic stability of (7.1), and hence we see that (7.1) is *globally asymptotically stable if and only* $r(A) < 1$. In this case we will say $A$ is *stable*. For computational criteria that the roots of a polynomial lie in the unit circle see [21]. Although we have assumed, and continue to do so, that $A$ is real, nothing we have done so far uses this assumption.

7.3. *Exercise.* Show that

$$w_1(n) = \lambda_1^n,$$

$$w_j(n+1) = \sum_{k=0}^{n} \lambda_j^{n-k} w_{j-1}(k), \quad j = 2, \ldots, m.$$

7.4. *Exercise.* If the eigenvalues of $A$ are distinct, show that

$$w_1(n) = \lambda_1^n,$$

$$w_j(n) = \sum_{i=1}^{j} c_{ij} \lambda_i^n,$$

where

$$c_{ij}^{-1} = \prod_{k=1,\, k \neq i}^{j} (\lambda_i - \lambda_k).$$

7.5. *Exercise* (Variation of constants formula). Show that the solution of the initial problem ($f : J_+ \to R^m$)

$$x' = Ax + f(n), x(0) = x^0,$$

is

$$x(n+1) = A^{n+1}x^0 + \sum_{k=0}^{n} A^{n-k}f(k).$$

7.6. *Exercise.* Show that the solution of

$$y(n+m) + a_1 y(n+m-1) + \cdots + a_m y(n) = g(n) \quad (7.5)$$

satisfying $y(0) = y(1) = \cdots = y(m-1) = 0$ is

$$y(n+1) = \sum_{k=0}^{n} w(n-k)g(k),$$

where $w$ is the $m$th principal solution of the homogeneous equation ($g(n) \equiv 0$); that is, $w$ is the solution of

$$y(n+m) + a_1 y(n+m-1) + \cdots + a_m y(n) = 0$$

satisfying

$$y(0) = \cdots = y(m-2) = 0, \qquad y(m-1) = 1.$$

Since $A^{k+1} - A^k = A^k(A - I)$, we see that

$$A^{n+1} - I = (A - I)\sum_{k=0}^{n} A^k.$$

If $A^{n+1} \to 0$ ($r(A) < 1$), we obtain

$$(I - A)^{-1} = \sum_{k=0}^{\infty} A^k. \quad (7.6)$$

$R(\lambda) = (\lambda I - A)^{-1}$ is called the *resolvent* of $A$. We see, therefore, if $|\lambda| > r(A)$ that

$$R(\lambda) = (\lambda I - A)^{-1} = \sum_{k=0}^{\infty} \lambda^{-(k+1)}A^k. \quad (7.7)$$

It is also of interest to know when it is true that each solution of (7.1) approaches a point—which, of course, must be an equilibrium point. Assume that $A^n x \to y = f(x)$ for each $x$ ($\Omega(x) = \{f(x)\}$). Since $f(x)$ is linear, $f(x) = Bx$ for some matrix $B$, and the question we are asking is: *when does $A^n$ converge*? We know, if $r(A) < 1$, that $A^n \to 0$ as $n \to \infty$, and, if $r(A) > 1$, the sequence is unbounded. This leaves the case $r(A) = 1$. Assume that $A^n \to B$ as $n \to \infty$ (i.e., $A^n x \to Bx$ for each $x$). Hence, the point $Bx$ is an equilibrium point of $\dot{x} = Ax$ (a fixed point of $A$) and $ABx = Bx$. Since $B \neq 0$, a necessary condition for $A^n \to B$ when $r(A) = 1$ is that $\lambda = 1$ be an eigenvalue of $A$ and be the only eigenvalue on the unit circle. Now $A^n \to B$ implies $A^n$ is bounded, and we see from equation 7.7 that $(\lambda - 1)R(\lambda)$ is bounded for $\lambda > 1$. Hence $\lambda = 1$ is a simple pole of the resolvent (this means that 1 is a simple root of the minimal polynomial). The converse of this is true (see, for instance, [37], Chapter I), and hence

7.7. THEOREM: *$A^n \to 0$ if and only if $r(A) < 1$. If $r(A) > 1$, $A^n$ is unbounded and does not converge. If $r(A) = 1$, then $A^n$ converges if and only if $\lambda = 1$ is a simple pole of the resolvent of $A$ and is the only eigenvalue of $A$ on the unit circle.*

This can also be seen from the algorithm for $A^n$ or from the Jordan canonical form for $A$.

7.8. *Exercise.* Show that: If $A^n \to B$ as $n \to \infty$, then $AB = BA = B$ and $B^2 = B$.

7.9. *Exercise.* When is it true that each solution of (7.1) is bounded?

## 8. STABILITY OF LINEAR SYSTEMS

To illustrate further the application of Liapunov's direct method and the use of Liapunov functions, we will study a bit more the question of the stability of

$$\dot{x} = Ax. \tag{8.1}$$

We shall say, in place of "the origin for (8.1) is asymptotically stable", simply, "(8.1) is asymptotically stable" or $A$ is stable. Also, for linear systems asymptotic stability is always global, and the adjective is not needed. We know that $A$ is stable if and only if $r(A) < 1$. The next criterion is the analog of the one given originally by Liapunov for the real parts of all the eigenvalues of $A$ to be negative ($e^{At} \to 0$ as $t \to \infty$).

Let $V(x) = x^T Bx$ where $B$ is positive definite. Then, with respect to (8.1), $\dot{V}(x) = x^T(A^T BA - B)x$. Hence, if $A^T BA - B$ is negative definite, (8.1) is asymptotically stable by Exercise 5.9, and $A$ is stable. Conversely, suppose that $A$ is stable, and consider the equation

$$A^T BA - B = -C. \tag{8.2}$$

If it has a solution, then

$$-\sum_{k=0}^{n} (A^T)^k CA^k = (A^T)^{n+1} BA^{n+1} - B.$$

Letting $n \to \infty$, we see that the solution must be

$$B = \sum_{k=0}^{n} (A^T)^k CA^k.$$

It is easily verified that this is a solution and that, if $C$ is positive definite, it is positive definite.

Hence, we have shown

8.1. THEOREM: *If there are positive definite matrices $B$ and $C$ satisfying* (8.2), *then $A$ is stable. Conversely, if $A$ is stable, then given $C$,* (8.2) *has a unique solution $B$. If $C$ is positive definite, $B$ is positive definite.*

This result plays an important role in the theory of linear discrete control systems. It is also a converse theorem. If (8.1) is asymptotically stable, there is a positive definite quadratic Liapunov function $V$ with $-\dot{V}$ positive definite. (An easy proof of the first statement in Exercise 8.3 is obtained using this result.)

8.2. Consider the nonlinear difference equation

$$x' = Ax + f(n, x). \tag{8.3}$$

Assume that $f(n, x)$ is $o(x)$ uniformly with respect to $n \geqslant 0$. This means that given $\varepsilon > 0$ there is a $\delta > 0$ such that $\|f(n, x)\| < \varepsilon\|x\|$ for all $\|x\| < \delta$ and all $n \geqslant 0$. Near the origin $f(n, x)$ should not have much effect on stability. To some extent this is true and that is the next exercise (see [28]).

8.3. *Exercise* (Stability by the linear approximation). Consider the nonlinear difference equation (8.3) where $f(n, x)$ is $o(x)$ as described above. Show that: If the linear approximation $x' = Ax$ is asymptotically stable ($r(A) < 1$), then so is (8.3). If $r(A) > 1$, then (8.3) is unstable.

The last statement in the result above is more difficult to prove. The critical cases are when $r(A) = 1$. Then one must take into account the nonlinearities. This result on the linear approximation is useful in applications. It was, up until 1950, the way most control systems were designed. But it must be kept in mind that it gives no information about the size of the region of asymptotic stability. It is purely a local result. The region of asymptotic stability may be so small relative to a given application that from a practical point of view the origin is unstable. Also the origin of the nonlinear system can be unstable but a very small neighborhood of the origin could be an attractor and from a practical point of view it could be stable. There could be a stable periodic oscillation about the origin so small that its effect could be negligible. One of the advantages of Liapunov's direct method versus deciding stability on the basis of the linear approximation is that it takes the nonlinearities into account and can yield information about the extent of the stability or instability.

Nonnegative matrices, $A = (a_{ij})$ are those for which $a_{ij} \geqslant 0$ ($A \geqslant 0$). They arise naturally in many applications, and have been studied extensively for a long time. (See [4, 11].) For instance, for the linear difference equation $x' = Ax$ where the state variables are naturally nonnegative (populations, prices, number of particles, etc.) the matrix $A$ will be nonnegative since $\overline{R^m_+} = \{x; x \geqslant 0, x \in R^m\}$ must be positively invariant.

8.4. *Exercise.* Show that:
(a) $\overline{R^m_+}$ is positively invariant for $x' = Ax$ if and only if $A \geqslant 0$;
(b) $R^m_+ = \{x; x > 0, x \in R^m\}$ is positively invariant for $x' = Ax$ if and only if $A \geqslant 0$ and no row of $A$ is zero.

A good illustration of the use of our results is the proof of part of the following theorem:

8.5. THEOREM: *If $A \geqslant 0$, the following are equivalent:*
(1) $|\lambda(A)| < 1$.
(2) $(A - I)^{-1} \leqslant 0$.
(3) *There is a $c > 0$ such that $Ac < c$.*
(4) *$(A^T - I)D + D(A - I)$ is negative definite for some positive definite diagonal matrix $D$.*
(5) *The real parts of the eigenvalues of $A - I$ are all negative.*
(6) *The principal minors of $I - A$ are all positive.*

*Proof:* We shall prove the equivalence of the first three statements. (1) $\Rightarrow$ (2) follows from Equation (7.6). To see that (2) $\Rightarrow$ (3) take $c = (I - A)^{-1}b$ for any $b > 0$. Then, since $(I - A)^{-1} \geqslant 0$ and is nonsingular, $c > 0$. Assume (3) and let $w_i(x) = \dfrac{|x_i|}{c_i}$. We will show that $w$ is a vector Liapunov function. Here $W(x) = \operatorname{Max}_i \dfrac{|x_i|}{c_i}$;

$$w_i(T(x)) = \frac{|(Ax)_i|}{c_i} \leqslant \frac{1}{c_i} \sum_{k=1}^{m} a_{ik} c_k \frac{|x_k|}{c_k} \leqslant W(x) \frac{(Ac)_i}{c_i}$$

$$< W(x), \ x \neq 0.$$

Hence, $\mathring{w}(x) < 0$ for $x \neq 0$ and $w(x)$ is a vector Liapunov function. Since $W(x) \to \infty$ as $\|x\| \to \infty$, every solution is bounded and approaches $M$, which is simply the origin. Therefore, (3) $\Rightarrow$ (1), and we have shown the equivalence of (1), (2), and (3).

It is known that (3), (4) and (5) are equivalent. A rather detailed discussion of such equivalences can be found in [11]. What we have seen is that our stability theory gives simple proofs of some significant and nontrivial results. For applications of results of this type to economics see [19].

8.6. *Exercise.* Define $|A| = (|a_{ij}|)$. With the vector Liapunov function $w$ used in the above proof, show that: $|A|$ stable implies $A$ is stable, and hence that $r(A) \leqslant r(|A|)$.

8.7. *Exercise.* Show that:
(a) $|Ax| \leqslant |A| \cdot |x|$, $(|x|_i = |x_i|)$.
(b) If $|T(x)| \leqslant B|x|$ for all $x$ and $B$ is stable, then $T^n x \to 0$ for all $x$.

## 9. STABILITY UNDER PERTURBATIONS

The converse theorems of Liapunov theory are important theoretically—too important not at least to mention. They guarantee the existence of Liapunov functions if there is asymptotic stability, as did our earlier result for linear systems (Theorem 8.1). The general results are nonconstructive and are of no help in finding Liapunov functions, but, as we shall see, they do enable us to answer an important practical question.

A proof of the following converse theorem can be found in [15]. Here we consider again the general difference equation

$$\dot{x} = T(x),\ T(0) = 0. \tag{9.1}$$

The system has an equilibrium which we locate at the origin. $T$ is said to be *lipschitz continuous* near the origin if for some $L > 0$ and $r > 0$

$$\|T(x) - T(y)\| \leqslant L\|x - y\|$$

for all $\|x\| < r$ and $\|y\| < r$.

9.1. THEOREM: *If $T$ is lipschitz continuous near the origin and the origin is asymptotically stable, there exists a positive definite*

*$V : R^m \rightarrow R$ which is lipschitz continuous near the origin with $\dot{V}$ negative definite.*

Now our definitions of stability and asymptotic stable, which are Liapunov's, are in terms of perturbations of initial conditions. In the real world, where nothing is known exactly, a system is being constantly perturbed and a better model for the perturbations is

$$x' = T(x) + P(n, x), \tag{9.2}$$

where $P(n, x)$ is unknown but hopefully not too large. Now simple stability of the origin for the unperturbed system (9.1) is too fragile to expect it to imply a stability of the perturbed system (9.2). Not so with asymptotic stability, and this is why asymptotic stability is of practical interest. We will now describe this stability under perturbations. It is called "strong" because originally (and this was for ordinary differential equations) only stability under perturbations was proved (for instance, this is all that is proved in [15] for difference equations). The stronger result was first obtained within the context of topological dynamics (see [38]).

9.2. DEFINITION: Let $x^*(n, x^0)$ denote the solution of (9.2) satisfying $x^*(0, x^0) = x^0$. The origin is said to be *stable under perturbations* if for some $\varepsilon_0$ and each $\varepsilon > 0$ there exist $\delta_1(\varepsilon)$ and $\delta_2(\varepsilon)$ such that $\|\dot{x}\| < \delta_1(\varepsilon)$ and $\|P(n, y)\| < \delta_2(\varepsilon)$ for all $n \geqslant 0$ and all $\|y\| < \varepsilon_0$ imply $\|x^*(n, x^0)\| < \varepsilon$ for all $n \geqslant 0$. If, in addition, there is a $\delta_0$, an $r_0$, and an $N(\varepsilon)$ such that $\|x\| < \delta_0$ and $\|P(n, y)\| \leqslant \delta_2(\varepsilon)$ for all $n \geqslant 0$ and all $\|y\| \leqslant r_0$ imply $\|x^*(n, x)\| < \varepsilon$ for all $n \geqslant N(\varepsilon)$, the origin is said to be *strongly stable under perturbations*.

The following theorem can then be proved using the above converse theorem:

9.3. THEOREM: *If $T$ is lipschitz continuous in a neighborhood of the origin, then the origin is strongly stable under perturbations if and only if it is asymptotically stable.*

A rather recent and significant development for ordinary differential equations is the introduction of skew-product flows (see [39]) and their use to establish invariance properties for the positive limit sets of solutions of nonautonomous equations (see [9, 10, 27]). This then gives for a large class of nonautonomous ordinary differential equations ($dx/dt = f(t, x)$) an invariance principle and a stability theory much like what we have developed here for autonomous systems. The same can be done for nonautonomous difference equations $x' = T(n, x)$ (see [28]). This is a new tool for the analysis of stability of nonautonomous systems that has not yet been exploited. Unfortunately, this goes beyond the scope of this article.

Although this theory of difference equations has been presented in a concise and sophisticated language, much of it can be made easily accessible to undergraduates, particularly, if one is willing to study first dimensions 2 and 3. Some of it would make a good introduction to the study of ordinary differential equations. Linear difference equations also provide a nontrivial application within an elementary course on linear algebra.

At this point the author wishes to confess that this has been the first time he has looked systematically at the subject of difference equations. What he has done is to do everything by analogy with the more highly developed theory for ordinary differential equations and difference-differential equations (functional differential equations). It has been surprising to him to discover new results and to find at this level so interesting a theory and so much of practical significance. He is now an advocate for considering difference equations as a prerequisite to the study of differential equations, control and stability theory, and the theory of systems. Interesting phenomena modeled by difference equations are not too difficult to find, computations are easy, and it is good introductory applied mathematics.

## REFERENCES

1. M. A. Aizerman and F. R. Gantmacher, *Absolute Stability of Regulator Systems*, Holden-Day, San Francisco, 1964.

2. K. Arrow, H. D. Block, and L. Hurwicz, "On the stability of competitive equilibrium II," *Econometrica*, **27** (1959), 82–109.
3. K. J. Arrow and F. H. Hahn, *General Competitive Analysis*, Holden-Day, San Francisco, 1971.
4. R. Bellman, *Introduction to Matrix Analysis*, McGraw-Hill, New York, 1960.
5. ———, "Vector Lyapunov functions," *SIAM J. Control*, **1** (1962), 32–34.
6. G. D. Birkhoff, "Dynamical Systems", *Amer. Math. Soc. Colloq. Publ. Vol.* **9**, Amer. Math. Soc., Providence, 1927.
7. Garrett Birkhoff and S. MacLane, *A Survey of Modern Algebra*, Macmillan, New York, 1947.
8. M. Cuénod and A. During, *A Discrete-Time Approach for System Analysis*, Academic Press, New York, 1969.
9. C. M. Dafermos, "An invariance principle for compact processes," *J. Differential Equations*, **9** (1971), 239–252.
10. ———, "Semiflows associated with compact and uniform processes," *Math. Systems Theory*, **8** (1974), 142–149.
11. M. Fiedler and V. Pták, "On matrices with nonpositive off-diagonal terms and positive principal minors," *Czechoslovak Math. J.*, **12** (1962), 382–400.
12. W. H. Fleming, *Functions of Several Variables*, Addison-Wesley, Reading, 1965.
13. W. H. Gottshalk and G. A. Hedlund, *Topological Dynamics*, Amer. Math. Soc., Providence, 1955.
14. Lj. T. Grujić and D. D. Šiljak, "Asymptotic stability and instability of large scale systems," *IEEE Trans. Automatic Control, AC*-**18** (1973), 636–645.
15. A. Halanay, "Quelques questions de la théorie de la stabilité pour les systèmes aux différences finies," *Arch. Rational Mech. Anal.*, **12** (1963), 150–154.
16. J. K. Hale, "Dynamical systems and stability," *J. Math. Anal. Appl.*, **26** (1969), 39–59.
17. M. W. Hirsch and S. Smale, *Differential Equations, Dynamical Systems, and Linear Algebra*, Academic Press, New York, 1974.
18. J. R. Hicks, *Value and Capital*, Oxford University (Clarendon) Press, London, 1939.
19. G. Horwich and P. A. Samuelson (Editors), *Trade, Stability, and Macroeconomics, Essays in Honor of Lloyd A. Metzler*, Academic Press, New York, 1974.
20. J. Hurt, "Some stability theorems for ordinary difference equations," *SIAM J. Numer. Anal.*, **4** (1967), 582–596.
21. E. I. Jury, *Inners and Stability of Dynamic Systems*, Wiley, New York, 1974.
22. R. E. Kalman and J. E. Bertram, "Control system analysis and design via the second method of Lyapounov II discrete systems," *Trans. ASME Ser. D., J. Basic Engineering*, **82** (1960), 394–400.
23. C. Kuratowski, *Topology I*, Academic Press, New York, 1966.

24. J. P. LaSalle, "The extent of asymptotic stability," *Proc. Nat. Acad. Sci. U.S.A.*, **46** (1960), 363–365.

25. J. P. LaSalle and S. Lefschetz, *Stability by Liapunov's Direct Method with Applications*, Academic Press, New York, 1961.

26. J. P. LaSalle, "Stability theory for ordinary differential equations," *J. Differential Equations*, **4** (1968), 57–65.

27. ———, "Stability theory and invariance principles," *Dynamical Systems: An International Symposium*, vol. I, editors L. Cesari, J. K. Hale, and J. P. LaSalle, Academic Press, New York, 1976.

28. ———, "Stability of dynamical systems," Conference Board of the Mathematical Sciences, Regional Conference Series in Mathematics, Mississippi State University, Aug. 1975. Published by SIAM, 1977.

29. S. Lefschetz, *Stability of Nonlinear Control Systems*, Academic Press, New York, 1965.

30. A. M. Liapunov, "Problème Général de la Stabilité du Mouvement," *Ann. of Math. Studies*, Princeton University Press, Princeton, 1947.

31. R. K. Miller and G. R. Sell, "Existence, uniqueness and continuity of solutions of integral equations," *Ann. Mat.*, **80** (1968), 135–152.

32. ———, "Topological dynamics and its relation to integral equations and nonautonomous systems," *Dynamical Systems: An International Symposium*, vol. I, editors L. Cesari, J. K. Hale, and J. P. LaSalle, Academic Press, New York, 1976.

33. W. E. Milne, *Numerical Calculus*, Princeton University Press, Princeton, 1949.

34. L. M. Milne-Thompson, *Calculus of Finite Differences*, Macmillan, London, 1933.

35. E. J. Putzer, "Avoiding the Jordan canonical form in the discussion of linear systems with constant coefficients," *Amer. Math. Monthly*, **73** (1966), 2–7.

36. J. E. Rubio, *The Theory of Linear Systems*, Academic Press, New York, 1971.

37. H. H. Schaefer, *Banach Lattices and Positive Operators*, Springer-Verlag, New York, 1974.

38. P. Seibert, *Stability under Perturbations in Generalized Dynamical Systems*, Nonlinear Differential Equations and Nonlinear Mechanics, editors J. P. LaSalle and S. Lefschetz, Academic Press, New York, 1963, 463–473.

39. G. R. Sell, "Nonautonomous differential equations and topological dynamics," *Trans. Amer. Math. Soc.*, **127** (1967), 241–283.

40. J. Stein, "Unemployment, inflation, and monetarism," *American Economic Review*, LXIV, (December 1974).

41. P. van den Driessche (Editor), "Mathematical problems in biology," *Victoria Conference, 1973*, Springer-Verlag, New York, 1974.

42. H. Uzawa, "The stability of dynamic processes," *Econometrica*, **29** (1961), 617–631.

43. T. Yoshizawa, "Stability theory by Liapunov's second method", *Publication No. 9*, The Mathematical Society of Japan, Tokyo, 1966.

# WHAT IS A DYNAMICAL SYSTEM?

*George R. Sell*

## I. HISTORICAL BACKGROUND

During the last quarter of the nineteenth century the study of ordinary differential equations underwent some rather radical changes. Prior to this time the major emphasis in the subject had been on the methods of "solving" various equations either by writing down specific formulae for the solutions or by expansion in terms of a series. It was during this generation that Peano (1890) used the polygonal method of Euler and Cauchy to give a rigorous proof of the existence theorem for solutions of ordinary differential equations. During the same period, Lipschitz (1876) and Picard (1890) showed that the method of successive approximations led to a proof of both the existence and uniqueness of solutions of the initial value problem, or as it is sometimes called, the Cauchy problem. These developments, which were to be refined by still other researchers in the years ahead, represent the denouement of the attempts to "solve" differential equations.[1]

At the same time as the works of Lipschitz, Peano and Picard were closing one chapter in the book on differential equations,

[1]See Hartman [10, p. 23] for more detailed references to these historical papers.

another was being opened with the research of Poincaré [20] and Lyapunov [12]. This new chapter, the study of the qualitative behavior of solutions, was based on an entirely different approach. In this case the existence of solutions is assumed, and one does not attempt to compute these solutions. Rather one tries to exploit the topological properties of the phase space and the analytical properties of the given vector field in order to determine the behavior of the solutions, say as time $t \to +\infty$.

During the fifty years following Poincaré many important advances were made in the qualitative theory of differential equations. As this was happening mathematicians began to realize that the essence of this new theory was based on the notion of a dynamical system. It took several years before the abstract formulation of a dynamical system was perfected. It is then with the work of G. D. Birkhoff [4] and V. V. Nemytskii and V. V. Stepanov [18] that we arrive at the stage where this paper begins.

## II. DEFINITION AND SOME EXAMPLES

We begin with a metric space $X$ and a metric $d$. A *dynamical system* or, as it is sometimes called, a *flow* is then a continuous mapping $\pi : X \times R \to X$, where $R$ denotes the real line, satisfying two properties:

(i) (Identity property) $\pi(x, 0) = x$ for all $x$ in $X$.

(ii) (Group property) $\pi(\pi(x, t), s) = \pi(x, t + s)$ for all $x$ in $X$ and $t, s$ in $R$.

Let us now look at several examples.

*Example* 1 (The Bebutov flow).

Let $C(R, R^n)$ denote the family of all continuous functions $h : R \to R^n$, where $R^n$ denotes the Euclidean $n$-space, with the topology of uniform convergence on compact sets. This topology is generated by the metric

$$d(g, h) = \sum_{n=1}^{\infty} 2^{-n} \frac{d_n(g, h)}{1 + d_n(g, h)},$$

where

$$d_n(g, h) = \max_{|t| \leqslant n} |g(t) - h(t)|.$$

For $h \in C(R, R^n)$, we define the translate $h_\tau$ by $h_\tau(t) = h(\tau + t)$. Then the mapping

$$\pi(h, \tau) = h_\tau$$

defines a flow on $C(R, R^n)$, cf. [24, pp. 29–32].

*Example* 2 (The parameterized Bebutov flow).

Let $W$ be an open subset[2] of $R^n$ and let $C(W \times R, R^n)$ denote the family of continuous functions $f : W \times R \to R^n$ with the topology of uniform convergence on compact sets. This topology is also metrizable, cf. [24, p. 36]. Once again if

$$f \in C(W \times R, R^n)$$

define the translate $f_\tau$ by

$$f_\tau(x, t) = f(x, \tau + t).$$

Then the mapping

$$\pi(f, \tau) = f_\tau$$

defines a flow on $C(W \times R, R^n)$.

We suggested above that the theory of dynamical systems originated with the study of differential equations. Let us now look more carefully at this connection.

*Example* 3 (Autonomous differential equations).

Let $W$ be an open set in $R^n$ and let $f : W \to R^n$ be a Lipschitz-continuous function; for example, $f$ may be a $C^1$-function. Then for any $x_0 \in W$, the initial value problem

$$x' = f(x), \quad x(0) = x_0, \tag{1}$$

admits a unique solution defined in some interval $-\alpha < t < \alpha$. Let us assume[3] for simplicity, that, for all $x_0 \in W$, the solution of

---

[2] $W$ can be more general in Example 2. It can be an arbitrary metric space, or even a topological space.

[3] This assumption is not necessary. In order to avoid it one would have to reformulate the definition of a dynamical system so as to allow flows where $\pi(x, t)$ is not defined for all $t$ in $R$, cf. [16, 24], for example.

(1) is defined for all $t$ in $R$, and let $\varphi(x_0, t)$ denote this solution. Then $\varphi(x_0, t)$ defines a flow on $W$. Indeed the fundamental theory of ordinary differential equations assures us that $\varphi : W \times R \to W$ is continuous, and $\varphi(x_0, 0) = x_0$ by definition. Moreover, since

$$\psi_1(t) = \varphi(\varphi(x_0, s), t) \quad \text{and} \quad \psi_2(t) = \varphi(x_0, s + t)$$

are both solutions of the initial value problem

$$x' = f(x), \quad x(0) = \varphi(x_0, s),$$

it follows from the uniqueness[4] of solutions that $\psi_1(t) = \psi_2(t)$ for all $t$ in $R$, i.e., the group property is satisfied.

*Example* 4 (Nonautonomous differential equations).

As in Example 3, we let $W$ denote an open set in $R^n$ and let $\mathfrak{F}$ be a collection of continuous functions $f : W \times R \to R^n$. We shall make the following two assumptions[5]:

(i) The collection $\mathfrak{F}$ is translation-invariant, i.e., $f_\tau \in \mathfrak{F}$, for all $\tau \in R$ whenever $f \in \mathfrak{F}$.

(ii) For each $x_0 \in W$ and $f \in \mathfrak{F}$ there exists a unique solution $\varphi(x_0, f, t)$ of the initial value problem

$$x' = f(x, t), \quad x(0) = x_0,$$

and, moreover, $\varphi(x_0, f, t)$ is defined for all $t$ in $R$.

The collection $\mathfrak{F}$ is a subset of $C(W \times R, R^n)$ with the topology of uniform convergence on compact sets. Under the above assumptions one can now show that the mapping

$$\pi(x, f, \tau) = (\varphi(x, f, \tau), f_\tau)$$

defines a flow[6] on $W \times \mathfrak{F}$, cf. [24, pp. 59–62].

Let us now turn from the examples to the objects we study in the theory of dynamical systems. Let $\pi$ be a given dynamical

[4]The assumption that $f$ be Lipschitz continuous, which guarantees the *uniqueness* of solutions of the initial value problem, can actually be dropped. However, this would be at the expense of constructing the flow generated by $x' = f(x)$ on a somewhat more complicated space, cf. [26] for details.

[5]These assumptions can be greatly relaxed, cf. [16, 24, 26].

[6]The main problem in showing that $\pi$ is a flow is in verifying that $\pi$ is continuous. This question of continuity is discussed in some detail in [16].

system on $X$. The *orbit* or *trajectory* through $x \in X$ is the point set

$$\gamma(x) = \{\pi(x, t) : t \in R\}.$$

The *positive* and *negative semi-trajectories* are the sets

$$\gamma^+(x) = \{\pi(x, t) : t \geqslant 0\} \quad \text{and} \quad \gamma^-(x) = \{\pi(x, t) : t \leqslant 0\}.$$

We shall also study the *hull*

$$H(x) = \text{Cl}\ \gamma(x)$$

as well as the *positive* and *negative hulls*

$$H^+(x) = \text{Cl}\ \gamma^+(x) \quad \text{and} \quad H^-(x) = \text{Cl}\ \gamma^-(x),$$

where Cl denotes the closure operation on $X$. In addition, there are the limit sets, the *alpha limit set*

$$A(x) = \bigcap_{\tau \in R} H^-(\pi(x, \tau)),$$

and the *omega limit set*

$$\Omega(x) = \bigcap_{\tau \in R} H^+(\pi(x, \tau)).$$

In our discussion of the historical background we saw that the study of dynamical systems was a natural development in the attempt to analyze the qualitative behavior of solutions of ordinary differential equations. Now we can be more specific about the philosophical questions underlying the theory of dynamical systems, *viz*.,

(i) What are the main results?

(ii) What are the primary tools or methods?

Since these very general questions are discussed in some detail in two survey papers, [16, 25], we shall focus our attention here on a more limited albeit very important aspect of the theory of dynamical systems, *viz*., the study of the structure of omega limit sets. Knowledge about the structure of the omega limit set $\Omega(x)$ leads one to a valuable insight into the asymptotic behavior, as $t \to +\infty$, of the motion $\pi(x, t)$. By studying several illustrative examples, we will now show how certain qualitative methods (especially topological methods) can be used to derive information about the structure of omega limit sets.

## III. ELEMENTARY PROPERTIES OF DYNAMICAL SYSTEMS

Let $\pi$ be a dynamical system on a metric space $X$. A subset $A \subseteq X$ is said to be an *invariant set* (or *positively invariant*) if $\gamma(x) \subseteq A$ (or $\gamma^+(x) \subseteq A$) whenever $x \in A$. One can easily show that if $A$ is invariant, then the closure $\mathrm{Cl}\, A$ is also invariant. It follows then that for every $x \in X$ the hull $H(x)$ is a closed invariant set. Also one can show that the omega limit set $\Omega(x)$ is a closed invariant set; however, it can happen that $\Omega(x)$ may be an empty set.

The function $t \to \pi(x, t)$ is called the *motion* through the point $x$. We say that the motion $\pi(x, t)$ is *positively compact* if the positive hull $H^+(x)$ is compact. This is equivalent to saying that there is a compact set $K \subseteq X$ with $\pi(x, t) \in K$ for all $t \geqslant 0$. In the case of a positively compact motion, one can derive some interesting facts about the omega limit set, cf. [24, pp. 20–21].

THEOREM 1: *Let $\pi(x, t)$ be a positively compact motion. Then the omega limit set $\Omega(x)$ is a nonempty, compact, connected, invariant set in $X$.*

A subset $M \subseteq X$ is said to be a *minimal set* if (i) $M$ is nonempty, closed and invariant, and (ii) $M$ has no proper subset with the above properties. By using the Nested Sequence Theorem for nonempty compact sets, one can easily show that every nonempty compact invariant set contains a minimal set. Therefore, if the motion $\pi(x, t)$ is positively compact, then $\Omega(x)$ contains a minimal set.

A fixed point and a periodic orbit are the simplest examples of minimal sets. A point $x \in X$ is a *fixed point* if $\pi(x, t) = x$ for all $t$. A point $x \in X$ is a *periodic point* (and the orbit $\gamma(x)$ is a *periodic orbit*) if there is an $\omega > 0$ such that $\pi(x, t) = \pi(x, t + \omega)$ for all $t$. (In the latter case we also say that the motion $\pi(x, t)$ is periodic.) More complicated examples of minimal sets are described in [18].

## IV. THE POINCARÉ–BENDIXSON THEORY

Let us now assume that the phase space $X$ is homeomorphic to a subset of the Euclidean plane $R^2$, and we let $\pi$ be a flow on $X$. Furthermore we shall assume that $\pi$ is generated by an autonomous differential equation $x' = f(x)$ where $f$ is a $C^2$-function.

The Poincaré–Bendixson theory gives a rather complete description of the omega limit sets which arise in this flow. This theory is a beautiful illustration of the use of topological techniques (specifically, the Jordan Curve Theorem) in the study of differential equations. We do not have the space here to derive this theory so we refer the reader to the discussion in [8, 10] for more details.

The first result describes the compact minimal sets.

THEOREM 2: *Let $M$ be a compact minimal set in $X$. Then either $M$ is a fixed point $\{x\}$ or $M$ is a periodic orbit $\gamma(x)$.*

The next result describes the structure of the omega limit sets. For our purposes we shall be primarily interested in the case where the omega limit set does not contain any fixed points.

THEOREM 3: *Assume that $\pi(x, t)$ is a positively compact motion with the property that $\Omega(x)$ contains no fixed points. Then $\Omega(x)$ is a periodic orbit.*

If it happens that $\pi(x, t)$ is positively compact and $\Omega(x)$ contains a finite number of fixed points, then one can conclude that either

(i) $\Omega(x)$ is a single fixed point, or

(ii) $\Omega(x)$ is homeomorphic to the circle $S^1$, and the flow in $\Omega(x)$ consists of fixed points and "transit orbits" connecting these fixed points, cf. [8].

In the next two sections we shall look at some applications of the Poincaré–Bendixson theory, but before we do that let us consider the following interesting consequence of Theorem 3:

Let $f : R \times R \to R$ be a $C^1$-function that is periodic in $t$, say that $f(x, t+1) = f(x, t)$ for all $x$ and $t$. Let $\mathfrak{F} = \{f_\tau : \tau \in R\}$ and assume that for each $x_0 \in R$ and $\tau \in R$ the initial value problem

$$x' = f_\tau(x, t), \quad x(0) = x_0,$$

admits a solution $\varphi(x_0, f_\tau; t)$ defined for all $t$ in $R$. Let

$$\pi(x, f, \tau) = (\varphi(x, f, \tau), f_\tau)$$

denote the flow generated on $R \times \mathfrak{F}$ as described in Example 4. Assume that for some $x_0 \in R$ and $\tau \in R$, the solution $\varphi(x_0, f_\tau, t)$ is bounded for all $t \geqslant 0$. Then one can conclude that the differential equation $x' = f(x, t)$ has a periodic solution. This conclusion, as we now show, is a direct application of Theorem 3. First, since $f(x, t)$ is periodic in $t$, the space $\mathfrak{F}$ is either a circle $S^1$ or a set consisting of a single point. In the latter case the function $f$ is independent of $t$, and the differential equation $x' = f(x)$ is autonomous. Let us exclude this case and assume that $f$ is not autonomous.[7] Then $R \times \mathfrak{F} = R \times S^1$ is homeomorphic to a subset of $R^2$. Next the flow $\pi$ has no fixed points. Since the motion $\pi(x_0, f_\tau, t)$ remains in a compact set in $R \times \mathfrak{F}$ for all $t \geqslant 0$, it follows from Theorem 3 that the omega limit set $\Omega(x_0, f_\tau)$ is a periodic orbit. Since $\Omega(x_0, f_\tau)$ contains a point of the form $(y, f)$, it follows that $\varphi(y, f, t)$ is a periodic solution of $x' = f(x, t)$.

## V. THE VAN DER POL EQUATION

We shall now use the Poincaré–Bendixson theory to show that the second-order differential equation

$$u'' + \varepsilon(u^2 - 1)u' + u = 0 \tag{2}$$

---

[7] In the case that $f$ is autonomous, one can show directly that, under our assumption, $x' = f(x)$ has a fixed point and, therefore, a periodic solution.

(where $\varepsilon$ is a positive constant) has a nontrivial periodic solution. (A fixed point is referred to as a trivial periodic solution.)

Our first step is to change (2) into a first order system in the plane $R^2$ by setting $x_1 = u$, $x_2 = u' = v$ and

$$x' = \begin{pmatrix} u \\ v \end{pmatrix}' = \begin{pmatrix} v \\ -\varepsilon(u^2 - 1)v - u \end{pmatrix} = f(x). \tag{3}$$

We see that $f$ is a $C^2$-function and one can show that for every $x_0 \in R^2$, the solution $\varphi(x_0, t)$ of the initial value problem

$$x' = f(x), \quad x(0) = x_0,$$

is defined for all $t$ in $R$. Since $f(x) = 0$ implies that $x = 0$, we see that the origin $x = 0$ is the only fixed point of the flow $\varphi$.

We shall now construct a Jordan curve $\Gamma$ in the plane $R^2$ with the property that every solution beginning in the interior Int $\Gamma$ remains in Int $\Gamma$ for all $t \geqslant 0$, cf. Figure 1. But before doing this let us make some preliminary observations.

For $R > 0$ we let $C_R$ denote the circle $V(u, v) = R^2$, where $V(u, v) = u^2 + v^2$. Now if $(u(t), v(t))$ denotes a solution of (3) and $V(t) = V(u(t), v(t))$ then

$$\frac{dV}{dt} = \text{grad } V \cdot f(x) = 2uu' + 2vv' = -2\varepsilon(u^2 - 1)v^2.$$

Since $dV/dt < 0$ when $u^2 > 1$ and $v \neq 0$, we see that for $R > 1$ no solution of (3) can exit from Int $C_R$ across the arcs of $C_R$ with $u \geqslant 1$ or $u \leqslant -1$.

Next let $U$ be the subset of the $uv$-plane defined by

$$U = \{(u, v) : 1 < u \quad \text{and} \quad \varepsilon(u^2 - 1)v < -u\}.$$

Then at any point $(u_0, v_0)$ in $U$ one has $v' > 0$.

With those observations we now define positive numbers $m$, $v_0$, $R$, $S$ and $u_2$ by $m = 2\varepsilon$, $v_0 = 6\varepsilon + 2\varepsilon^{-1}$, $R^2 = v_0^2 + 1$,

$$S^2 = (v_0 + 2m)^2 + 1$$

and

$$u_2^2 = S^2 - (v_0 - 3m)^2.$$

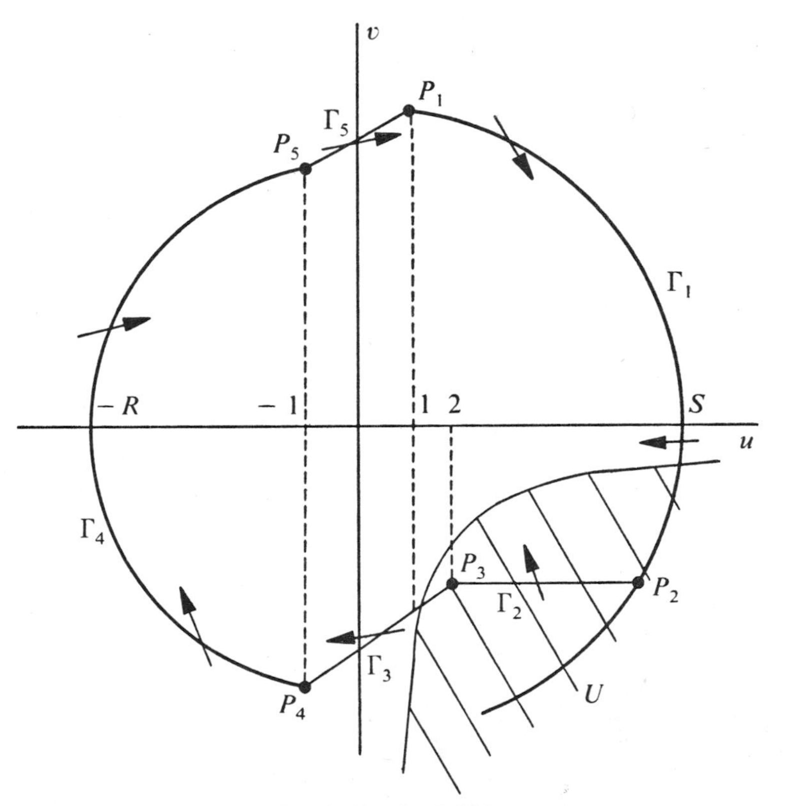


FIG. 1. Van der Pol Flow

Then define the points $P_1, \ldots, P_5$ by

$$P_1 = (1, v_0 + 2m), \quad P_2 = (u_2, -v_0 + 3m),$$

$$P_3 = (2, -v_0 + 3m), \quad P_4 = (-1, -v_0), \quad P_5 = (-1, v_0).$$

The points $P_1$ and $P_2$ lie on $C_S$, and $P_4$ and $P_5$ lie on $C_R$. Also $u_2 > 2$, therefore $P_3$ lies to the left of $P_2$. Furthermore $P_3$ is in the region $U$ and therefore any point on the straight line $v = -v_0 + 3m$ to the right of $P_3$ also lies in $U$, i.e., the line segment $[P_3P_2]$ lies in $U$. Define the Jordan curve $\Gamma = \Gamma_1 \cup \Gamma_2 \cup \Gamma_3 \cup \Gamma_4 \cup \Gamma_5$ as follows:

(i) $\Gamma_1$ = the arc $[P_1P_2]$ of $C_S$ with $u \geqslant 1$.
(ii) $\Gamma_2$ = the line segment $[P_2P_3]$.
(iii) $\Gamma_3$ = the line segment $[P_3P_4]$.
(iv) $\Gamma_4$ = the arc $[P_4P_5]$ of $C_R$ with $u \leqslant -1$.
(v) $\Gamma_5$ = the line segment $[P_5P_1]$.

The observations made above show that no solution of (3) can exit from Int $\Gamma$ across the curve $\Gamma_1$ or $\Gamma_4$. Since $\Gamma_2 \subseteq U$ and $v' > 0$ on $\Gamma_2$ we see that no solution of (3) can exit from Int $\Gamma$ across $\Gamma_2$.

The line segments $\Gamma_3$ and $\Gamma_5$ each have slope $= m$. If we can show that $dv/du \leqslant m$ along each of these segments, then (since $u' > 0$ when $v > 0$ and $u' < 0$ when $v < 0$) it follows that no solution of (3) can exit from Int $\Gamma$ across $\Gamma_3$ or $\Gamma_5$. Now

$$\frac{dv}{du} = -\varepsilon(u^2 - 1) - \frac{u}{v}.$$

Therefore on $\Gamma_3$ one has

$$\frac{dv}{du} \leqslant \varepsilon - \frac{u}{v} \leqslant \varepsilon - \frac{2}{-v_0 + 3m}.$$

It is easy to verify that the right hand term is $= m$. Similarly on $\Gamma_5$ one has

$$\frac{dv}{du} \leqslant \varepsilon - \frac{u}{v} \leqslant \varepsilon + \frac{1}{v_0},$$

and the right hand term is $< m$.

Now fix any $x_1 \in \Gamma_2$. Then the solution $\pi(x_1, t)$ remains in the interior of $\Gamma$ for all $t \geqslant 0$. Thus $\pi(x_1, t)$ is positively compact. In order to complete the application of Theorem 3 to (3) we must show that the omega limit set $\Omega(x_1)$ contains no fixed points. For this purpose we shall use the following linearization theorem[8].

THEOREM 4: *Assume that $W$ is an open set in $R^2$ and that $f: W \to R^2$ is a $C^2$-function. Let $x_0 \in W$ be a fixed point of the*

[8] The linearization theorem, which we formulate for differential equations in the plane $R^2$, admits an extension to higher dimensions, cf. [5, 8, 10].

*differential equation* $x' = f(x)$, *i.e.*, $f(x_0) = 0$. *Let* $A = f_x(x_0)$ *denote the Jacobian matrix of* $f$ *evaluated at* $x_0$. *Then the following statements are valid*:

(i) *If* $\operatorname{tr} A > 0$ *and* $\det A > 0$, *then* $x_0$ *is a source for* $x' = f(x)$, *i.e., there is a neighborhood* $U$ *of* $x_0$ *with the property that for all* $x \in U - \{x_0\}$ *one has* $A(x) = \{x_0\}$ *and* $\pi(x, t)$ *remains outside* $U$ *for all* $t \geqslant \tau_x$, *for some* $\tau_x > 0$.

(ii) *If* $\operatorname{tr} A < 0$ *and* $\det A > 0$, *then* $x_0$ *is a sink for* $x' = f(x)$, *i.e., there is a neighborhood* $U$ *of* $x_0$ *with the property that for all* $x \in U - \{x_0\}$ *one has* $\Omega(x) = \{x_0\}$ *and* $\pi(x, t)$ *remains outside* $U$ *for all* $t \leqslant \tau_x$, *for some* $\tau_x < 0$.

(iii) *If* $\det A < 0$, *then* $x_0$ *is a saddle*[9] *for* $x' = f(x)$, *i.e., there is a neighborhood* $U$ *of* $x_0$ *with the property that the two sets*

$$\{x \in U : \Omega(x) = \{x_0\}\} \quad \text{and} \quad \{x \in U : A(x) = \{x_0\}\}$$

*are one-dimensional manifolds.*

If we apply this theorem to the van der Pol equation (3), we see that the Jacobian matrix at the origin (0, 0) is

$$A = \begin{pmatrix} 0 & 1 \\ -1 & \varepsilon \end{pmatrix}.$$

Since $\det A = 1$ and $\operatorname{tr} A = \varepsilon > 0$ we see that the origin is a source for equation (3), i.e., the given trajectory $\pi(x_1, t)$ does not enter a neighborhood $U$ of the origin. Since the origin is the only fixed point for (3), we see that $\Omega(x_1)$ contains no fixed points. It now follows from Theorem 3 that the van der Pol equation (3) has a nontrivial periodic solution for any $\varepsilon > 0$.

Let us now turn to a problem arising from mathematical biology.

## VI. POPULATION DYNAMICS IN A FOOD CHAIN

Picture now a food chain with three agents. Let $x_1$, $x_2$ and $x_3$ denote the population of each agent, and assume that $x_1$ is food

[9] The precise definition of a saddle can be found in [8, pp. 106–117]. We have used here a property of a saddle which is needed later.

for $x_2$ and that $x_2$ is food for $x_3$. This type of food chain arises in the phenomena of phagocytosis and bacteriophage. We shall assume that $x_1$ is being added to the system at a rate $r > 0$ and that all three agents are being discharged from the system at a rate proportional to the respective population. The equations of growth have the form

$$\begin{aligned} x_1' &= r(1 - x_1) - x_2 f_1(x_1) \\ x_2' &= -rx_2 + x_2 f_1(x_1) - x_3 f_2(x_2) \\ x_3' &= -rx_3 + x_3 f_2(x_2) \end{aligned} \tag{4}$$

where $f_i(\xi) = k_i \xi (a_i + b_i \xi)^{-1}$, for $i = 1, 2$ and the coefficients $a_1, a_2, b_1, b_2, k_1, k_2$ are positive. Under these conditions it is easy to show that the solutions of (4) are defined for all $t$ in $R$.

Now set $y = x_1 + x_2 + x_3$. Then $y' = r(1 - y)$ and therefore $y(t) \to 1$ as $t \to +\infty$. Consequently, in order to study the limiting behavior of (4) as $t \to +\infty$, it will suffice to study the limiting surface $y = 1$. In this way, the 3-dimensional problem given by (4) becomes a 2-dimensional problem, which we can write in terms of the $x_1 x_2$-coordinates as

$$\begin{aligned} x_1' &= r(1 - x_1) - x_2 f_1(x_1), \\ x_2' &= -rx_2 + x_2 f_1(x_1) - (1 - x_1 - x_2) f_2(x_2). \end{aligned} \tag{5}$$

We hope now to be able to apply the Poincaré–Bendixson theory to (5). Our objective is to find conditions on the seven parameters $r, a_1, a_2, b_1, b_2, k_1, k_2$ which guarantee the existence of a nontrivial periodic solution.

Since $x_1$, $x_2$ and $x_3$ represent population sizes, it is reasonable then to restrict our attention to the set

$$K = \{(x_1, x_2, x_3) : x_i \geqslant 0 \quad \text{for} \quad i = 1, 2, 3 \quad \text{and} \quad x_1 + x_2 + x_3 = 1\}.$$

The following result gives a mathematical justification of this biological "fact":

THEOREM 5: *The set $K$ is positively invariant, i.e., if $P_0 = (x_{10}, x_{20}, x_{30}) \in K$, then the solution through $P_0$ remains in $K$ for all $t \geq 0$.*

*Proof:* It will suffice to look at the boundary of $K$ in the surface $y = 1$. At $x_1 = 0$ we have $x_1' = r > 0$. Also at $x_2 = 0$ we have $x_2' = 0$, and at $x_3 = 0$ we have $x_3' = 0$. Therefore the solution through $P_0$ cannot leave $K$ at any time $t \geq 0$.

Let us now search for the fixed points of (4) in the surface $y = 1$. We see that there are only three, viz.,

(I) $x_1 = 1, \quad x_2 = x_3 = 0.$

(II) $r = f_1(x_1), \quad x_1 + x_2 = 1, \quad x_3 = 0.$

(III) $r = f_2(x_2) = x_2 f_1(x_1)(1 - x_1)^{-1}, \quad x_1 + x_2 + x_3 = 1.$

The first singularity remains in $K$ for all $r > 0$. However the second and third singularities can be outside of $K$ when $r$ is large. For example, if $r > f_1(1)$ then clearly the second singularity is not in $K$. In order to be more precise about this, we define the number $r_0$ by

$$r_0 = f_1(x_1) = f_2(x_2), \quad x_1 + x_2 = 1,$$

and we let $r_1 = f_1(1)$. One can show that $r_0$ is well-defined and that the following theorem is valid:

THEOREM 6: *One has $0 < r_0 < r_1$. Furthermore the following statements are valid:*

(A) *If $0 < r < r_0$, then all three singularities lie in $K$ and the third singularity is in the interior of $K$.*

(B) *If $r_0 < r < r_1$, then only the first two singularities lie in $K$.*

(C) *If $r_1 < r$, then only the first singularity lies in $K$.*

We can now apply the linearization theorem (Theorem 4) to the second order system (5) and determine whether each of the three singularities is a saddle, a sink or a source. We then have the results described in Table I.

TABLE I. Singularities for (5)

| Singularity | $0 < r < r_0$ | $r_0 < r < r_1$ | $r_1 < r$ |
|---|---|---|---|
| I | saddle | saddle | sink |
| II | saddle | sink | not in $K$ |
| III | see below | not in $K$ | not in $K$ |

As we have seen, the third singularity lies in $K$ when one has $0 < r < r_0$. In this case, the Jacobian matrix $A$ at the third singularity satisfies $\det A > 0$. Also the formula for the trace is given by

$$\operatorname{tr} A = -r + f_1(x_1) - x_2 f_1'(x_1) - x_3 f_2'(x_2), \tag{6}$$

where $x_1$, $x_2$ and $x_3$ are evaluated at the third singularity (III) and $f_i'(\xi) = a_i k_i (a_i + b_i \xi)^{-2}$ for $i = 1, 2$. The linearization theorem then informs us that (III) is a source if $\operatorname{tr} A > 0$ and a sink if $\operatorname{tr} A < 0$. Now, in the event that (III) is a source, then (III) cannot lie in the omega limit set of any other point $P \in K$. However since (I) and (II) are saddle points, it follows from the characterization of a saddle in Theorem 4 that there is a two-dimensional set of points $P$ in $K$ with the property that $\Omega(P)$ does not contain either (I) or (II). Therefore there is at least one point $P \in K$ with the property that $\Omega(P)$ contains no fixed points. We have thus proved the following result:

THEOREM 7: *Assume that $r$ is given so that $0 < r < r_0$ and $\operatorname{tr} A > 0$, where $\operatorname{tr} A$ is given by (6). Then (4) has a nontrivial periodic orbit in $K$.*

By a straightforward, but very laborious computation, one can show that if $r = 0.1$ and

$$\begin{aligned} a_1 &= 1, \quad b_1 = 0.1, \quad k_1 = 1, \\ a_2 &= 2, \quad b_2 = 1, \quad k_2 = 1, \end{aligned}$$

then $r < r_0$ and the third singularity (III) is a source.

The behavior of solutions of (4) for $r > r_0$ is illustrated in

Figure 2. The behavior of solutions of (4) for $0 < r < r_0$, where (III) is a source[10], is illustrated in Figure 3.

## VII. CONCLUDING REMARKS

The Poincaré–Bendixson theory and the related structure of omega limit sets comprises just a small corner of the general qualitative theory of dynamical systems. Research into the latter theory is extensive and is concerned with many diverse problems, some of which are discussed elsewhere in this volume. The reader who is interested in further study may find the following outline helpful:

1. Flows for more general systems: Oftentimes one is faced with a model which does not fit into the examples discussed in Section

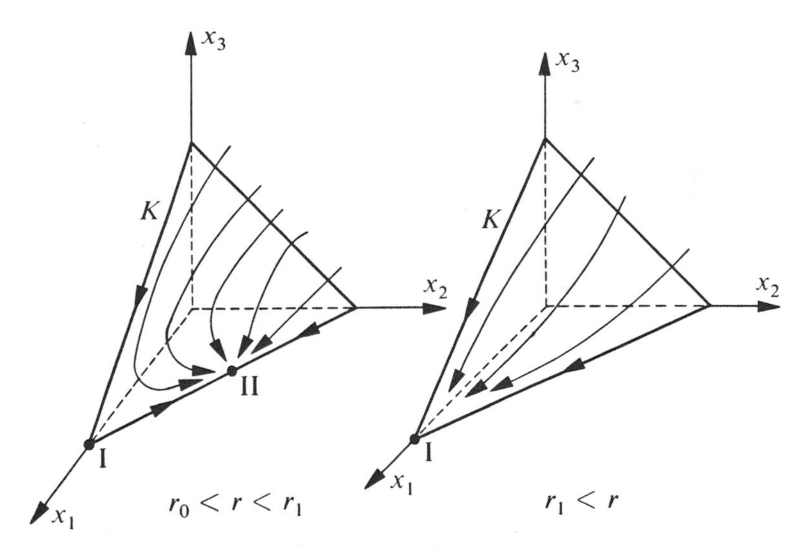


FIG. 2. Food Chain Flow

[10]The case where the third singularity is a sink is unresolved. We do not know whether or not there exists a nontrivial periodic orbit in this case.

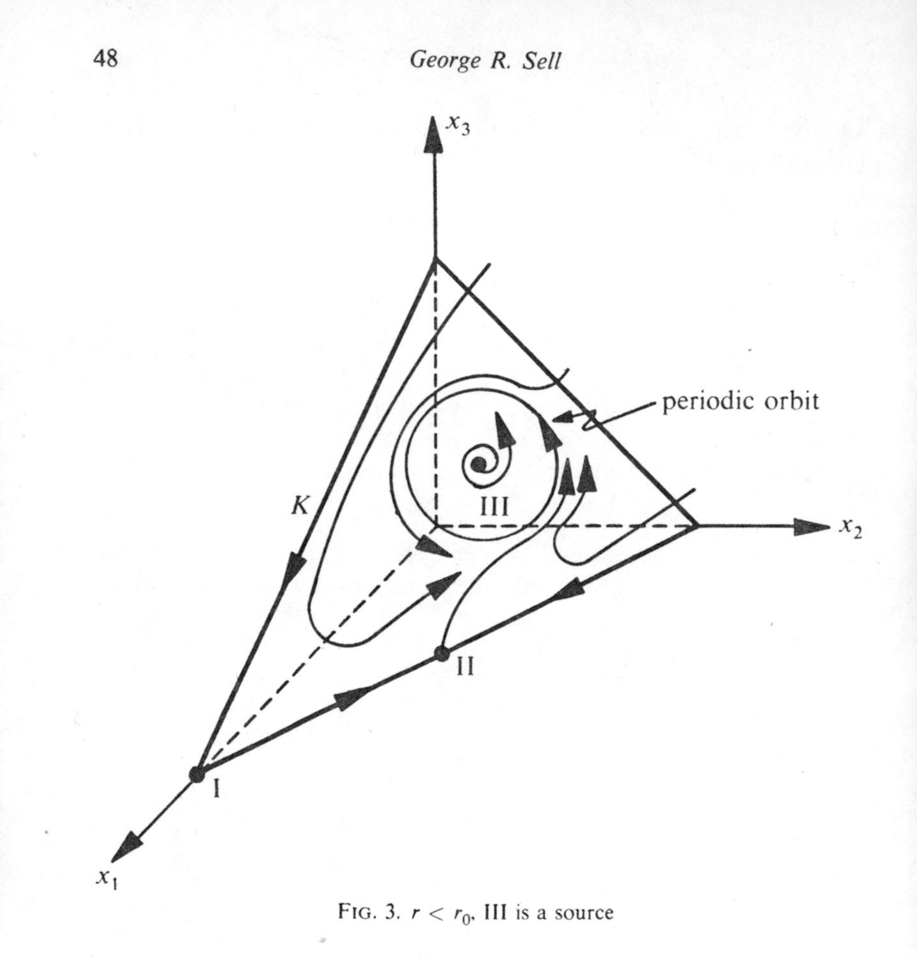


FIG. 3. $r < r_0$, III is a source

II. For example, one may have a system with discontinuous coefficients, or one may have a Volterra integral system which is not reducible to a simple differential equation. In order to use the techniques of topological dynamics in these cases, one needs to construct a flow of greater generality than treated here. A general introduction to this problem can be found in [2, 16, 25, 26].

2. Stability theory: Stability theory, in itself, is a very broad subject and permeates many aspects of the qualitative theory of differential equations. An introduction to Lyapunov stability theory can be found in [3, 11]. Structural stability and the related concept of differentiable dynamics are discussed in [19, 29].

3. Existence of almost periodic solutions: The concept of almost periodicity is closely related to a plethora of mathematical disciplines. Probably for this reason, one finds that this topic is one of the most widely studied areas in the qualitative theory of differential equations. Recent accounts of the theory of almost periodic solutions can be found in [7, 17, 21, 27, 30].

4. Hamiltonian systems and celestial mechanics: The general theory of differential equations began with such physical problems as falling bodies and planetary motion. It is not surprising then that the qualitative theory should be strongly influenced by many of the same problems. As it happens, these problems fall into the general area of Hamiltonian systems. A general introduction into the dynamics of Hamiltonian systems, along with a brief historical discussion, can be found in [1, 13, 17]. A special example of interest is the motion of a satellite in the gravitational field of the solar system. This system is modelled with a nonautonomous differential equation. Control-theoretic problems such as space-rescue missions can now be formulated in this context, and the theory of dynamical systems is a fundamental tool for the analysis of such problems, cf. [14, 15].

5. Linear vs. Nonlinear theories: To a great extent the qualitative theory of differential equations is a theory of techniques or methods. One of the techniques which is widely employed is the method of linearization. The idea here is that linear equations are more tractable than nonlinear equations, and one hopes that the dynamical behavior of solutions of the linearized equation will lead to insight into the behavior of solutions of the original equation, which is in general nonlinear. An expository account of these theories can be found in [9] and in the forthcoming book [22]. Also see [28] for some recent advances in this area.

We would like to express our sincere appreciation to Professor Hans Weinberger for his helpful comments and suggestions on the food chain problem.

REFERENCES

1. V. I. Arnold and A. Avez, *Ergodic Problems of Classical Mechanics*, Benjamin, New York, 1968.

2. Z. Artstein, "Topological dynamics of ordinary differential equations and Kurzweil equations," *J. Differential Equations*, **23** (1977), 224–243.

3. N. P. Bhatia and G. P. Szegö, *Stability Theory of Dynamical Systems*, Springer-Verlag, New York, 1970.

4. G. D. Birkhoff, *Dynamical Systems*, Amer. Math. Soc. Colloq. Publ., Providence, 1927.

5. E. A. Coddington and N. Levinson, *Theory of Ordinary Differential Equations*, McGraw-Hill, New York, 1955.

6. J. Cronin, "Periodic solutions in $n$ dimensions and Volterra equations," *J. Differential Equations*, **19** (1975), 21–35.

7. A. M. Fink, *Almost Periodic Differential Equations*, Lecture Notes in Math. 377, Springer-Verlag, New York, 1974.

8. J. K. Hale, *Ordinary Differential Equations*, Wiley-Interscience, New York, 1969.

9. J. K. Hale and J. P. LaSalle, "Differential equations: Linearity vs. Nonlinearity," *SIAM Rev.*, **5** (1963), 249–272.

10. P. Hartman, *Ordinary Differential Equations*, Wiley, New York, 1964.

11. J. P. LaSalle, "Invariance principles and stability theory for nonautonomous systems," *Proc. Greek Math. Soc.*, Carathéodory Symposium, Athens, 1973.

12. A. M. Lyapunov, *Problème Général de la Stabilité du Mouvement*, Ann. Math. Studies 17, Princeton University Press, Princeton, 1947.

13. L. Markus and K. R. Meyer, "Generic Hamiltonian dynamical systems are neither integrable nor ergodic," *Mem. Amer. Math. Soc.* 144, Providence, 1974.

14. L. Markus and G. R. Sell, "Capture and control in conservative dynamical systems," *Arch. Rational Mech. Anal.*, **31** (1968), 271–287.

15. ———, "Control in conservative dynamical systems: Recurrence and capture in aperiodic fields," *J. Differential Equations*, **16** (1974), 472–505.

16. R. K. Miller and G. R. Sell, "Topological dynamics and its relations to integral equations and nonautonomous systems," *Dynamical Systems: An International Symposium*, vol. I, Academic Press, New York, 1976, 223–249.

17. J. Moser, *Stable and Random Motions in Dynamical Systems*, Ann. Math. Studies 77, Princeton University Press, Princeton, 1973.

18. V. V. Nemytskii and V. V. Stepanov, *Qualitative Theory of Differential Equations*, Princeton University Press, Princeton, 1960.

19. Z. Nitecki, *Differentiable Dynamics*, MIT Press, Cambridge, 1971.

20. H. Poincaré, "Mémoire sur les courbes définies par une équation differentielle, I, II, III et IV," *J. Math. Pures Appl.*, (3) **7** (1881), 375–422; (3) **8** (1882), 251–286; (4) **1** (1885), 167–244; and (4) **2** (1886), 151–217.

21. R. J. Sacker and G. R. Sell, "Lifting properties in skew-product flows with applications to differential equations," *Mem. Amer. Math. Soc.*, (to appear)

22. ———, *Linear Differential Systems*, (to appear).

23. G. R. Sell, "Periodic solutions and asymptotic stability," *J. Differential Equations*, **2** (1966), 143–157.

24. ———, *Topological Dynamics and Ordinary Differential Equations*, Van Nostrand Reinhold, London, 1971.

25. ———, "Topological dynamical techniques for differential and integral equations," *Ordinary Differential Equations*, Academic Press, New York, 1972, 287–304.

26. ———, "Differential equations without uniqueness and classical topological dynamics," *J. Differential Equations*, **14** (1973), 42–56.

27. ———, "A book review," *Bull. Amer. Math. Soc.*, **82** (1976), 198–207.

28. ———, "The structure of a flow in the vicinity of an almost periodic motion", *J. Differential Equations*, (to appear).

29. S. Smale, "Differentiable dynamical systems," *Bull. Amer. Math. Soc.*, **73** (1967), 747–817.

30. T. Yoshizawa, *Stability Theory and the Existence of Periodic and Almost Periodic Solutions*, Appl. Math. Sciences 14, Springer-Verlag, New York, 1974.

# GENERIC PROPERTIES OF ORDINARY DIFFERENTIAL EQUATIONS

*M. M. Peixoto*

## 1. GENERIC THEORY AND STRUCTURAL STABILITY

### 1.0 Introduction

We present here an elementary introduction to the *generic theory* of autonomous, ordinary differential equations. This theory, part of the *qualitative or geometrical theory* of these equations, studies the qualitative features that are present in "most equations", in the following sense. We put a topological structure on a given family of differential equations. Since it is not feasible to analyze the qualitative properties of every one of these differential equations, we restrict our analysis to the properties of an appropriate dense subset of this family. The concept of *structural stability*, to be explained below, is the basic general concept behind this theory.

It was through the generic theory that the ideas and methods of differential topology began to influence, in a systematic way, the theory of ordinary differential equations. Credit for this should be given to S. Smale and R. Thom, among others. Nowadays the natural place to consider the global properties of ordinary differential equations is on a differentiable manifold. In this setting, a differential equation is called a "flow" or "dynamical system". A differentiable manifold is the concept in differential topology that generalizes the usual concepts of curves and surfaces of the Euclidean space.

However, some of the more important concepts of the generic theory are of a local character and can be explained without previous knowledge of differential topology. Also, it turns out that in the development of the generic theory on manifolds, no knowledge of the "geometry" of the manifold is normally required (the exception is the two dimensional case). Further, the "differentiable structure" of the manifolds can be ignored. These facts serve to validate our decision to introduce the subject on Euclidean space, where we can easily convey the main features of the generic theory. Even then, we will have a lot of material to cover and only a few proofs will be given.

We shall consider only differential equations defined on the unit disc $D^n : x_1^2 + \cdots + x_n^2 \leqslant 1, n \geqslant 2$. Although many results that we will mention are usually stated for differential equations defined on a compact differentiable manifold, it usually makes no difference if we assume that this manifold is actually $D^n$. This should be kept in mind when we give references to the literature. Also, some of these results are phrased for "diffeomorphisms" instead of differential equations, but in the cases we shall consider there is basically no difference. To prove one of the basic results of the generic theory—the theorem of Kupka–Smale—one has to use Thom's transversality lemma from differential topology. We shall just give an *ad hoc* explanation of this lemma.

In this first chapter, we sketch the development of the qualitative theory and explain how the main ideas of the generic theory evolved. In doing so we give some definitions and introduce some necessary terminology. We assume the reader is acquainted with the basic properties of ordinary differential equations. A good reference book is Hirsch and Smale [1]. The reader is also referred to Chapters 13 and 16 of Coddington and Levinson [2].

## 1.1 The qualitative theory

Poincaré, in his famous mémoire of 1881 "Sur les courbes définies par une équation différentielle" [3], was the first to look at an ordinary differential equation from the point of view of the geometry of the set of trajectories, thereby establishing the geometric or qualitative theory. The Poincaré–Bendixson theorem illustrates this point of view.

The original motivation of Poincaré for introducing the qualitative theory came from problems in celestial mechanics. He used these problems, as well as the classical theory stemming from the work of Riemann on functions of a complex variable, to justify his work in the new field of topology (analysis situs, he called it). G. D. Birkhoff's [4] contributions to the qualitative theory early in this century were also motivated by problems in celestial mechanics. The stability theory created by Liapunov in his 1892 thesis [5], although heavily analytical in its methods, stressed the importance of some qualitative features of ordinary differential equations.

In spite of these outstanding classical contributions, in the 1950's the qualitative theory of Poincaré and Birkhoff still lacked the structure that was needed to incorporate it into the main stream of modern mathematics. In particular,

(i) Neither Poincaré nor Birkhoff ever made clear what was meant by saying that two differential equations were qualitatively equivalent, and

(ii) No topological structure had ever been given to a family of differential equations.

## 1.2 The equivalence relation

The pioneering work of Kneser in 1924 [6] and of Kaplan in 1941 [7] led to a formal definition of the (topological) equivalence of two differential equations. They needed this equivalence to classify certain differential equations.

To simplify matters we assume that we have just two differential equations

$$\dot{x}_1 = dx_1/dt = X_1(x_1, x_2), \quad \dot{x}_2 = dx_2/dt = X_2(x_1, x_2) \quad (1.2.1)$$

and

$$\dot{x}_1 = Y_1(x_1, x_2), \quad \dot{x}_2 = Y_2(x_1, x_2) \quad (1.2.2)$$

defined on the unit disc

$$D^2 : x_1^2 + x_2^2 \leqslant 1$$

and such that

> the vectors $(X_1, X_2)$ and $(Y_1, Y_2)$ are always transversal to the boundary of $D^2$ and point to the interior of $D^2$. (1.2.3)

Denote the differential equations (1.2.1) and (1.2.2) by their vector fields $X = (X_1, X_2)$ and $Y = (Y_1, Y_2)$, respectively. We always assume that our differential equations are of class $C^r$, $r \geqslant 1$, i.e., the components $X_1, X_2$ are functions of class $C^r$. The condition (1.2.3), that the vector fields are transversal to the boundary, is very convenient since we want the domain where our differential equations are defined to be compact. It also avoids certain complications at the boundary.

1.2.4 DEFINITION: The differential equations $X$ and $Y$ are *topologically equivalent*, denoted $X \sim Y$, if there is a homeomorphism

$$h : D^2 \to D^2$$

which maps trajectories of $X$ onto the trajectories of $Y$, and preserves the direction of the trajectories.

In particular, $h$ maps a singular point of $X$ into a singular point of $Y$, and every singular point of $Y$ is the image of a singular point of $X$. Likewise, $h$ maps the closed orbits of $X$ onto the closed orbits of $Y$. The set of all trajectories of a differential equation is sometimes called the phase portrait of $X$. When $X \sim Y$, we say that $X$ and $Y$ have the same phase portrait, the terminology that Kneser and Kaplan used.

### 1.3 Structural stability

Andronov and Pontrjagin [8] also introduced the idea of topological equivalence but indirectly, through the concept of *structural stability*.

We consider, as above, differential equations defined on the unit disc $D^2$, satisfying the transversality condition (1.2.3).

1.3.1 DEFINITION: The differential equation $X = (X_1, X_2)$ is said to be *structurally stable* ("grossière", they call it) if given $\varepsilon > 0$ there is $\delta > 0$ such that whenever the differential equation $Y = (Y_1, Y_2)$ is such that

$$|X_i - Y_i| < \delta, \quad \left| \frac{\partial (X_i - Y_i)}{\partial x_j} \right| < \delta, \quad i, j = 1, 2$$

at all points $(x_1, x_2) \in D^2$, then there is an $\varepsilon$-homeomorphism

$$h : D^2 \to D^2,$$

i.e., a homeomorphism moving each point by less than $\varepsilon$, which transforms trajectories of $X$ onto trajectories of $Y$ and preserves the direction of the trajectories.

In other words, $X$ is structurally stable when given $\varepsilon > 0$, then for $X$ and $Y$ close enough, $Y \sim X$ and the associated homeomorphism is an $\varepsilon$-homeomorphism. Andronov–Pontrjagin showed that structural stability in a differential equation was characterized by a very simple geometrical structure. To explain this we give a few definitions and results; see [1] for more details.

A *singularity* of the differential equation $X = (X_1, X_2)$ is a point $(x_1, x_2)$ where $X_1$ and $X_2$ both vanish. It is called *hyperbolic* if the eigenvalues of the corresponding Jacobian matrix have non-zero real part. There are only three possibilities for the qualitative behavior of a hyperbolic singularity: *sink*, *saddle point* and *source* (Figure 1). The number of eigenvalues with negative real part is 2, 1, and 0, respectively. For the moment, we are giving only a rough description of 2-dimensional hyperbolic singularities. We will say more about them later, but for the moment we observe that in (a) of Figure 1 we have two equivalent sinks and in (c) two equivalent sources.

We now consider a closed orbit $\gamma$ of the differential equation $X$ and through a point $0 \in \gamma$ pass a small segment $\sigma$ transversal to $X = (X_1, X_2)$ (see Figure 2). Then, following the trajectories of $X$ through points $x$ in $\sigma$ we get a map, the Poincaré transformation,

$$\pi : \sigma \to \sigma$$

defined for $x \in \sigma$ sufficiently close to 0.

The closed orbit $\gamma$ is said to be *hyperbolic* if $\pi'(0) \neq 1$. This condition is independent of the point 0 in $\gamma$. When $\pi'(0) < 1$, the

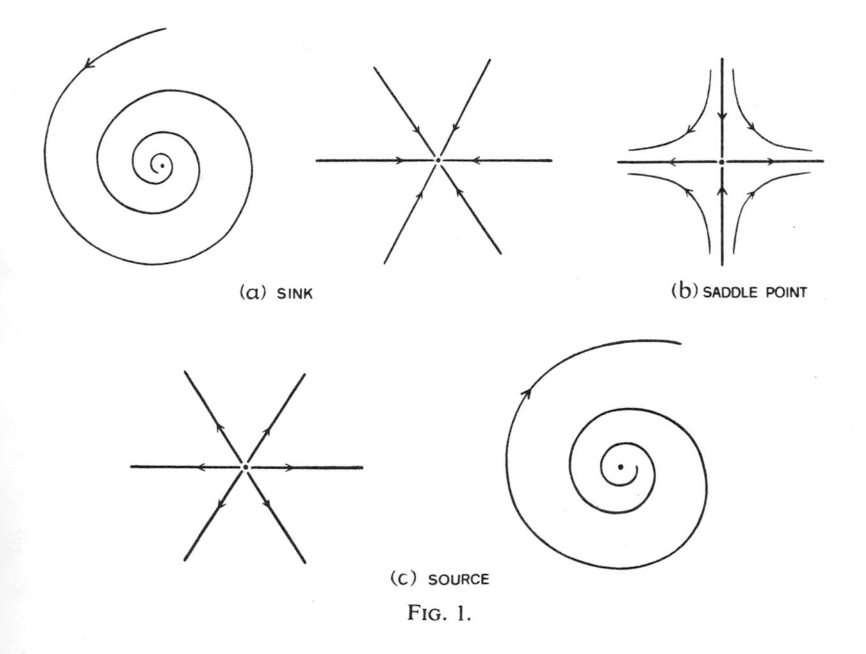


FIG. 1.

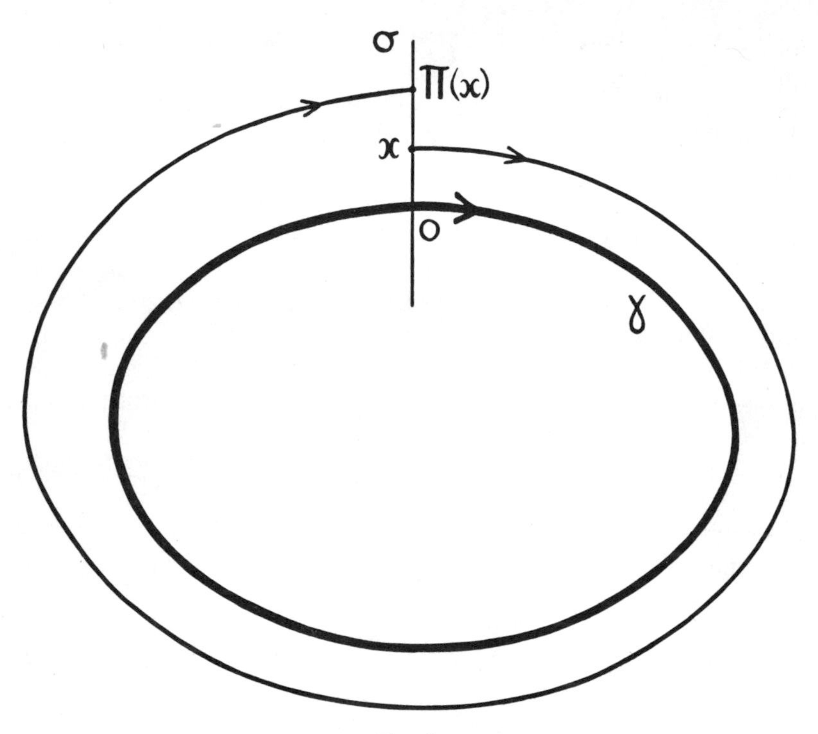


FIG. 2.

trajectories near $\gamma$ are spiraling strongly towards $\gamma$ and $\gamma$ is a stable limit cycle; when $\pi'(0) > 1$, they are spiraling strongly away from $\gamma$ and $\gamma$ is an unstable limit cycle. If all trajectories near $\gamma$ are closed, then $\pi'(0) = 1$. We can now state the characterization of Andronov–Pontrjagin.

1.3.2 THEOREM: *A necessary and sufficient condition in order that the differential equation $X$ be structurally stable is that the following three conditions be satisfied*:

(a) *the singularities of $X$ are hyperbolic*,

(b) *the closed orbits of $X$ are hyperbolic*, *and*

(c) *there is no trajectory of $X$ connecting saddle points*.

We will discuss this theorem later but for the moment we

observe that condition (a) implies that there is only a finite number of singularities and that conditions (b) and (c) together imply that there is only a finite number of closed orbits. Thus, structurally stable differential equations do have a very simple geometrical structure.

In Figure 3a we have the phase portrait of a differential equation $X$ which is structurally stable and in Figure 3b of another differential equation $Y$ which is not. We are assuming that the three singularities and the closed orbit of $X$ are all hyperbolic. The differential equation $Y$ violates condition (c) of Theorem 1.3.2 six times: each saddle point is connected with the other two and with itself.

Observe that through the phase portrait of a differential equation it is sometimes possible to tell that it is not structurally stable. However, it is never possible to guarantee that a differential equation is structurally stable just by looking at its phase portrait. For instance, assuming that the origin is one of the singularities of a structurally stable $X$, the differential equation $X'$ obtained by multiplying both components of $X$ by the factor $(x_1^2 + x_2^2)$ has exactly the same phase portrait as $X$ but $X'$ fails to be structurally stable. In fact, at the origin the singularity of $X'$ is not hyperbolic because the linear part has disappeared and both eigenvalues are zero.

## 1.4 The space of differential equations

It was in the context of structural stability that the construction of a space whose points are differential equations became natural and indeed necessary. This functional approach was first adopted in [9], as follows:

For the set $\mathfrak{X}$ of all differential equations defined on the disc $D^2$ and satisfying the transversality condition (1.2.3), define a distance between two differential equations $X = (X_1, X_2)$ and $Y = (Y_1, Y_2)$ by

$$d(X, Y) = \sup_{D^2}\left(|X_i - Y_i|, \left|\frac{\partial}{\partial x_j}(X_i - Y_i)\right|\right), \quad i, j = 1, 2. \qquad (1.4.1)$$

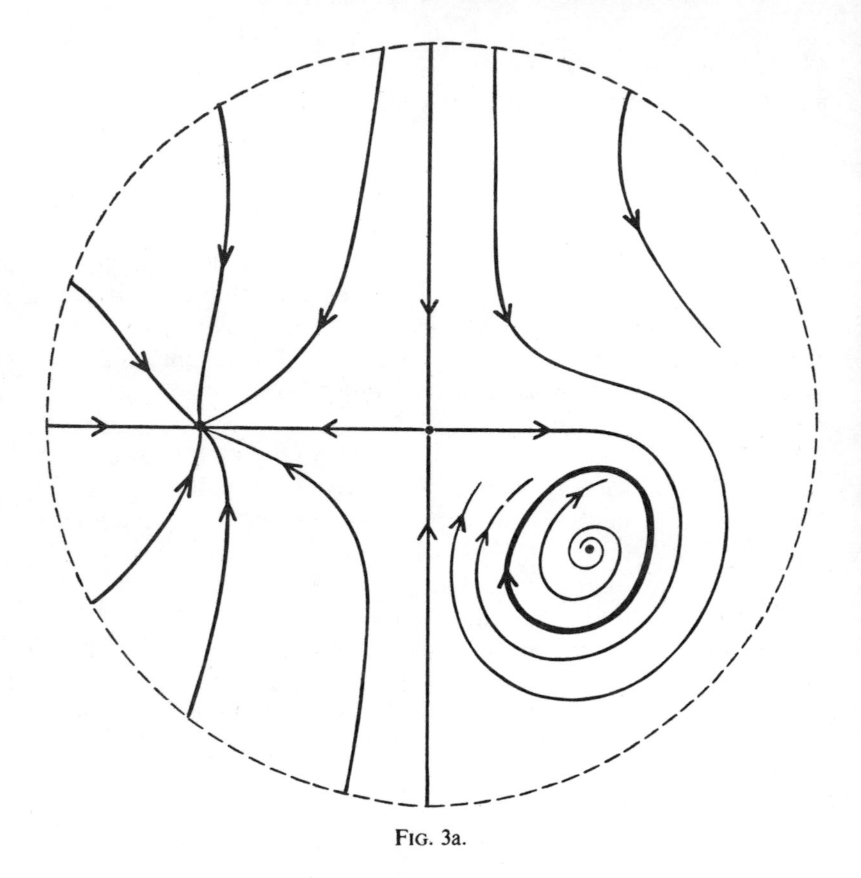

FIG. 3a.

Then $(\mathfrak{X}, d)$ is a metric space. This is called the $C^1$-topology on $\mathfrak{X}$. If $\Sigma$ is the subset of all structurally stable differential equations in $\mathfrak{X}$, then one can prove [9]:

1.4.2 THEOREM: *$\Sigma$ is open and dense in $\mathfrak{X}$.*

In other words, "almost all" differential equations are structurally stable.

In the process of proving this theorem, it was also found that the $\varepsilon$ is superfluous in the Andronov–Pontrjagin definition of structural stability, i.e., it is equivalent to the following.

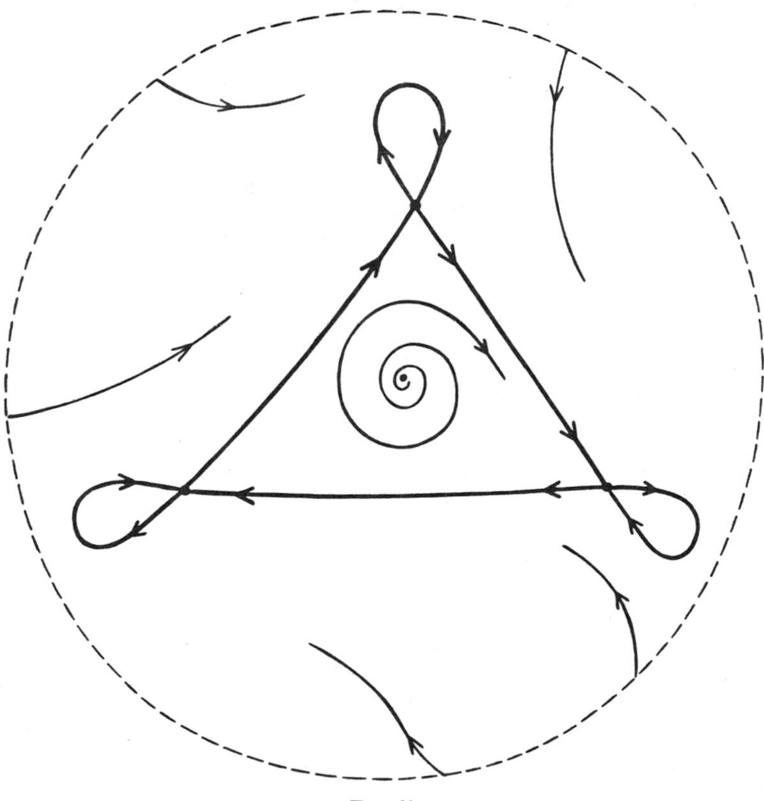

Fig. 3b.

1.4.3 Definition: $X \in \mathfrak{X}$ is structurally stable if and only if there is a neighborhood $\mathfrak{U}$ of $X$ in $\mathfrak{X}$ such that for all $Y \in \mathfrak{U}$, $Y \sim X$.

With this definition, the set $\Sigma$ of structurally stable differential equations is automatically open, whereas this was not the case with the $\varepsilon$-definition. Nowadays, 1.4.3 is the usual definition of structural stability. Another way to express the above definition is this: $X \in \Sigma$ if and only if $X$ is an interior point of one of the equivalence classes into which $\sim$ divides $\mathfrak{X}$.

## 1.5 Two comments about the definition of structural stability

Let us take a moment to explain the asymmetry of the fact that in $\mathfrak{X}$ we have the $C^1$-topology whereas $h : D^2 \to D^2$ is a homeomorphism which may not be differentiable. Because of this asymmetry, $h$ does not transform one differential equation into another. It would if it were differentiable, i.e., if it were a change of coordinates. In some questions this lack of differentiability is a source of trouble. On the other hand, there are compelling reasons to do as Andronov–Pontrjagin did, and this is what we intend to show.

We first note that from the point of view of structural stability it would not be appropriate to consider the $C^0$-topology so that in (1.4.1) only the differences $X_i - Y_i$ but not their derivatives would appear. For consider a differentiable function $y = f(x)$ having an isolated zero at $x = 0$. Then given $\varepsilon > 0$, we can find a differentiable function $\delta(x)$, whose $C^0$-norm is less than $\varepsilon$, i.e., $|\delta(x)| < \varepsilon$, and such that in a neighborhood $[-\alpha, \alpha]$ of the origin (depending on $\varepsilon$) $f(x) = -\delta(x)$. Now the function

$$f(x) + \delta(x)$$

is a $C^0$-perturbation of $f(x)$ of size less than $\varepsilon$ and it vanishes identically on $[-\alpha, \alpha]$, whereas $x = 0$ is an isolated zero of $f(x)$. This argument can be extended to functions of two variables where it can be applied to show that an arbitrarily small $C^0$-perturbation of a differential equation having an isolated singularity can change it into another equation exhibiting an open set of singularities.

A similar argument with the Poincaré transformation of a closed orbit can be used to show that an arbitrary small $C^0$-perturbation of a differential equation exhibiting an isolated closed orbit can change it into another differential equation exhibiting a band of closed orbits.

The above arguments show that $C^0$-perturbations are too strong to deal with singularities and closed orbits. Therefore, if we considered the $C^0$-topology in the definition of structural stability, no differential equation exhibiting singularities or closed orbits

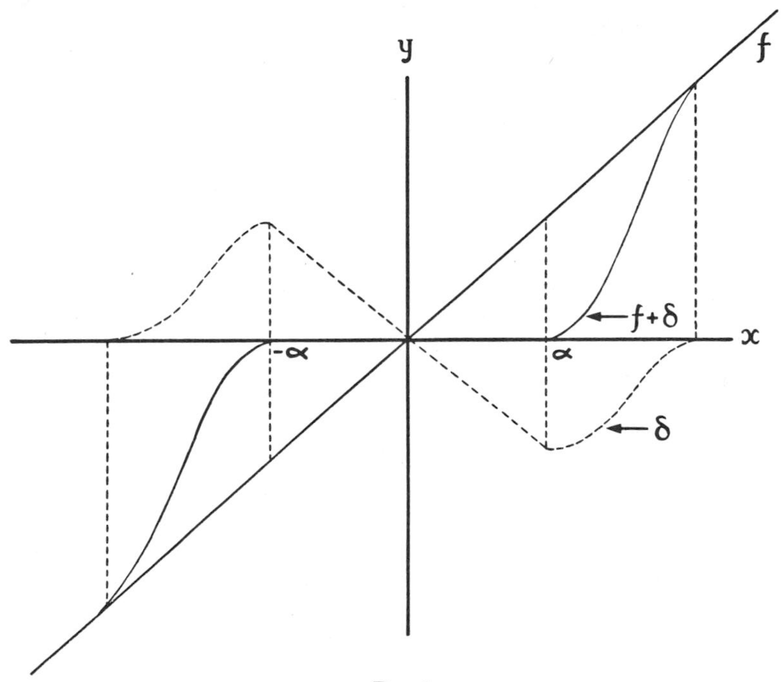


FIG. 4.

could be structurally stable. So the $C^1$-topology in $\mathfrak{X}$ is a necessary one.

Now that we know that $\mathfrak{X}$ must have the $C^1$-topology, it is natural to ask why the homeomorphism $h : D^2 \to D^2$ should not also be in class $C^1$. The answer is that if we did this, the eigenvalues at a singularity of $X$ and at the corresponding singularity of $Y$ would have to be proportional. Since one can always make a perturbation changing one eigenvalue and not the other, no singularity could possibly fit into a definition of structural stability if $h$ were of class $C^1$.

Let us illustrate this point about the eigenvalues at corresponding singularities of $X$ and $Y$. Without loss of generality we may assume that the origin $0 = (0, 0)$ is the common singularity of $X$ and $Y$ and that $h$, of class $C^1$, leaves the origin fixed, so that $h$ can be expressed as

$$y = Px + p(x), \quad y = (y_1, y_2), \quad x = (x_1, x_2), \qquad (1.5.1)$$

where $P$ is a $2 \times 2$ non-singular matrix and $p$ is of class $C^1$ and such that $p(0) = 0$, $dp(0) = 0$.

Now, in a neighborhood of the origin, the differential equation $X$ can be written

$$\frac{dx}{dt} = \dot{x} = Ax + a(x) \qquad (1.5.2)$$

with $A$ non-singular and $a(0) = 0$, $da(0) = 0$. Our assumption is that $h$, i.e., (1.5.1), maps trajectories of $X$ onto trajectories of $Y$. But (1.5.1) is actually a "change of coordinates" which can also be written as

$$x = P^{-1}y + q(y), \qquad (1.5.3)$$

with $q$ of class $C^1$ and $q(0) = 0$, $dq(0) = 0$. Differentiating (1.5.1) with respect to the time $t$ and taking into account (1.5.2) and (1.5.3) we see that the change of coordinates $h$ transforms (1.5.2), i.e., $X$, into the differential equation $Z$, given by

$$\dot{y} = PAP^{-1}y + o(y), \qquad (1.5.4)$$

where $o$ is a continuous function such that $o(y)/|y| \to 0$ as $|y| \to 0$. We remark that although $Z$ is not necessarily a differential equation of class $C^1$ everywhere, at the origin its Jacobian matrix is well defined and is equal to $PAP^{-1}$. Therefore, $X$ and $Z$ have the same eigenvalues at the origin.

Now, by construction, $h$ transforms trajectories of $X$ onto trajectories of $Z$, and since $h$ also transforms trajectories of $X$ onto trajectories of $Y$, we conclude that both $Y$ and $Z$ have the same set of trajectories. A further simple argument by Peixoto in [9] shows that $Y = KZ$ where the function $K = K(y)$ is continuous and $K(0) \neq 0$. But at the origin the Jacobian matrix of $Y$ is $K(0)$ times the Jacobian matrix of $Z$ and so the corresponding eigenvalues are proportional. Therefore the eigenvalues of $X$ and $Y$ at the origin are also proportional, as we wanted to show.

## 1.6 The generic theory

We have seen in sections 1.3 and 1.4 that the set $\Sigma$ of structurally stable differential equations is open and dense in the set $\mathfrak{X}$ of all differential equations, i.e., almost all differential equations are structurally stable, and that the trajectories of equations in $\Sigma$ have a very simple geometric structure. Moreover, the equivalence classes of $\Sigma$ modulo $\sim$ can be completely classified by means of graphs of a certain type [10]. Such a classification of $\Sigma$ modulo $\sim$ is the most one could hope for, as the following example demonstrates:

Let $K$ be a compact set consisting of the circle of radius $1/2$ centered at the origin together with an arbitrarily chosen compact set contained inside the disc $D_1^2$ bounded by this circle. From a classical result of analysis, one knows that there exists a $C^\infty$-function $\phi(x_1, x_2)$—a bump function—such that

$$\phi(x_1, x_2) = 0 \quad \text{if} \quad (x_1, x_2) \in K, \quad \text{and}$$

$$\phi(x_1, x_2) \neq 0 \quad \text{if} \quad (x_1, x_2) \notin K,$$

i.e., the set of zeros of $\phi(x_1, x_2)$ is precisely $K$. Now consider a differential equation $X = (X_1, X_2)$ such that between $D^2$ and $D_1^2$ the vector $(X_1, X_2)$ points to the origin along the radius and its length varies linearly from 1 at the boundary of $D^2$ to 0 at the boundary of $D_1^2$. Inside $D_1^2$ define $X$ by

$$\dot{x}_1 = \phi(x_1, x_2),$$

$$\dot{x}_2 = \phi(x_1, x_2).$$

Then the set of singularities of $X$ is the union of $K$ and the circle of radius $1/2$, and its trajectories are radii between the discs $D^2$ and $D_1^2$. Therefore, by letting the compact set $K$ vary inside $D_1^2$, we see that the classification of the corresponding differential equations modulo $\sim$ involves the classification modulo $\sim$ of all compact subsets of a disc. This problem is quite unmanageable and hopeless. Thus we cannot ask for more than a classification of the equations in the dense subset $\Sigma$. We can, however, attempt to extend to $D^n$ ($n > 2$) the nice theory we have for structurally stable differential equations on $D^2$.

Before we indicate how the theory generalizes to $D^n$ we remark that in $n$ dimensions we are considering differential equations

$$\dot{x}_i = X_i(x_1, \ldots, x_n), \quad i = 1, \ldots, n$$

defined on the unit disc $D^n$ and such that at the boundary of $D^n$ the vector $X = (X_1, \ldots, X_n)$ always points to the interior of $D^n$. The $C^1$-topology in the space $\mathfrak{X}$ of all such differential equations is generated by $d(X, Y)$, the $n$-dimensional generalization of the metric given by (1.4.1) (where now $i, j = 1, 2, \ldots, n$). Sometimes it is convenient to consider the $C^r$-topology, $r \geqslant 1$, in $\mathfrak{X}$ where the metric takes into account all derivatives of order up to $r$:

$$d_r(X, Y) = \sup_i |d^r(X_i - Y_i)|, \quad i = 1, \ldots, n.$$

In some cases, due to some technical reasons, it is necessary to do things with $\mathfrak{X} = \mathfrak{X}^r$ endowed with this $C^r$-topology, $r \geqslant 1$. In dimension 2, both the characterization theorem, Theorem 1.3.2, of Andronov-Pontrjagin and the density theorem, Theorem 1.4.2, are valid in $\mathfrak{X}^r$, $r \geqslant 1$. Unless we say otherwise, we shall henceforth assume that $\mathfrak{X}$ has a $C^r$-topology. With this topology, $\mathfrak{X}$ is a complete metric space.

The topological equivalence $\sim$ between elements $X, Y \in \mathfrak{X}$ is defined exactly as in case $n = 2$: there should exist a homeomorphism $h : D^n \to D^n$ mapping oriented orbits of $X$ onto oriented orbits of $Y$. The definition of structural stability in $n$-dimensions offers no difficulty: one takes the definition 1.4.3 which makes $\Sigma$, the set of all structurally stable differential equations, automatically open. Of course it would make sense to consider the $\varepsilon$-definition 1.3.1 but to this day it is not known whether or not for $n \geqslant 3$ both definitions are equivalent. Very likely they are. In all known examples of structurally stable differential equations the homeomorphism that one actually constructs can be made arbitrarily small. The comments of 1.5 remain valid here.

There are numerous examples of structurally stable differential equations in high dimensions. The most simple and best known are the Morse–Smale differential equations [22] which exhibit only a finite number of singularities and closed orbits. Others, like the Anosov differential equations [12], have a very complicated geometrical structure due to the fact that they possess infinitely many closed orbits.

In extending to $n$-dimensions our 2-dimensional theory for classifying a dense subset of differential equations, we are naturally led to a *generic theory* of differential equations and to the formulation of its *fundamental problem*. Properties like (a), (b), (c) of 1.3.2, i.e., the hyperbolicity of singularities and closed orbits and the nonexistence of saddle connections, that are true for an open and dense set of differential equations, are said to be *generic*. One would like to extend these three properties for dimension $n > 2$ and also to find other generic properties. Hopefully, after a number of these are found, one could classify modulo $\sim$ the differential equations in an open and dense set in $\mathfrak{X}$.

Condition (a) of 1.3.2 can be generalized in an obvious way to dimension $n > 2$ (eigenvalues have non-zero real part) and we get an open dense subset of $\mathfrak{X}$ whose singularities are hyperbolic. Conditions (b) and (c) can also be generalized but they will be satisfied on a dense, but not open, subset of $\mathfrak{X}$. This subset is a Baire set, that is, a set which is the countable intersection of sets open and dense in $\mathfrak{X}$.

We will discuss these particular generalizations further in the next sections. For the moment, we point out that while the formulations of conditions (a) and (b) in $n$-dimensions are fairly straightforward, the formulation of (c) is not quite obvious. It was done in 1960 by Smale [11] by introducing the ideas of "transversality" that were being used by Thom in differential topology. Smale's work was significant in that it marked the beginning of the contacts between the fields of differential equations and differential topology.

The above considerations suggest the following:

1.6.1 Definition: A property relative to an ordinary differential equation defined on $D^n$, $n \geqslant 2$, is said to be *generic* if and only if it is satisfied by all differential equations belonging to a Baire set $\Delta \subset \mathfrak{X}$.

Recall from general topology that a Baire set is dense in $\mathfrak{X}$ and that the intersection of two Baire sets is again a Baire set. Thus two Baire sets have a dense intersection. This is not the case for two dense sets, e.g., the rationals and irrationals on the line.

We now come to the formulation of the *fundamental problem* of generic theory. Its two parts are:

(i) finding a Baire set $\Delta \subset \mathfrak{X}$—the *approximation problem*, and

(ii) classifying $\Delta$ modulo the topological equivalence $\sim$—the *classification problem*.

The reader should be now asking: what about structural stability, the concept that triggered all these developments? Is structural stability a generic property? In 1966 Smale [13] gave a negative answer to this "problem of structural stability", showing that in dimension $n > 3$ the structurally stable differential equations are not dense in the space $\mathfrak{X}$ of all differential equations. Subsequently, Williams [14] extended this result to the case $n = 3$. So structural stability is a generic property only in case $n = 2$.

Part (i) of the fundamental problem, the approximation problem, has no nice solution. We will consider here the Baire set corresponding to conditions (a), (b), (c) of 1.3.2. For these conditions, the theorem of Kupka–Smale [15, 16] says that the density theorem 1.4.2 generalizes to dimension $n > 2$. Much work remains to be done on the approximation problem. Part (ii) of the fundamental problem, the classification problem, is no easier. Even taking for $\Delta$ something that is far from dense, the classification modulo $\sim$ of the Morse–Smale differential equations is a formidable problem if $n > 2$. All this suggests that to make any progress one should considerably restrict the scope of the generic theory. One way of doing this is to consider the *local generic theory*.

The local problem can be completely solved as we shall see in the next section. Even in the local situation we have to take the generic point of view, since the argument at the beginning of 1.6 shows that it is hopeless to classify all singularities modulo $\sim$.

Another way of having a more reasonable formulation of the fundamental problem is to relax the equivalence relation $\sim$ [17]. The problem then would consist of:

(i) finding an equivalence relation $E$ on $\mathfrak{X}$ which preserves as much of the orbit structure as possible,

(ii) finding a Baire set $\Delta$ in $\mathfrak{X}$; and

(iii) classifying the equivalence classes of $\Delta$ modulo $E$.

We say that one is "doing generic theory" whenever one finds some open set $O$ in $\mathfrak{X}$ and gets a Baire set $B \subseteq O$ of differential equations with nice geometrical properties which one could try to classify modulo some equivalence relation close to topological equivalence. In the next chapters we will discuss some examples of such generic theory. Let us close this chapter with a final remark about structural stability. In spite of the fact that structural stability is not a generic property, it turns out that all generic properties that have been found do exhibit a kind of structural stability, so that this concept still permeates generic theory. For instance, the non-density example of Smale in dimension $n$ is based on the fact that in dimension $n - 1$ a certain differential equation is structurally stable. Also structural stability has a strong physical appeal due to the fact that the equations governing natural phenomena are known only approximately. Broadly understood as the preservation of "form" under perturbation, the concept of structural stability is at the root of Thom's "catastrophe theory" [18].

## 2. LOCAL GENERIC THEORY

### 2.1 The fundamental theorem of the local theory

We consider, as before, differential equations defined on $D^n$ and call $\mathfrak{X}$ the space of all such equations with the $C^r$-topology, $r \geqslant 1$.

2.1.1 DEFINITION: Let $X, Y \in \mathfrak{X}$ and $p, q \in D^n$. We say that $X$ at $p$ is *locally equivalent* to $Y$ at $q$ if there are neighborhoods $V$ of $p$ and $W$ of $q$ such that there is a homeomorphism $h : V \to W$ mapping oriented arcs of trajectories of $X$ in $V$ onto the oriented arcs of trajectories of $Y$ in $W$.

Note that this definition would be valid if $X$ and $Y$ were defined only on these neighborhoods.

The following theorem settles the fundamental problem of the local generic theory. It says that, generically, there are only $n + 1$ possible types of local behavior for the trajectories of a differential equation in $D^n$.

2.1.2 THEOREM: *There is an open dense set $\mathcal{H} \subset \mathfrak{X}$ such that given $X \in \mathcal{H}$ and $p \in D^n$, then at $p$, $X$ is locally equivalent, at the origin, to one of the following $n + 1$ differential equations*:

$$\begin{aligned} &\dot{x}_1 = 1, \quad \dot{x}_j = 0, \quad j = 2, \ldots, n, \quad \text{or} \\ &\dot{x}_1 = -x_i, i \leqslant j, \quad \dot{x}_i = x_i, i > j, \quad j = 1, \ldots, n. \end{aligned} \tag{2.1.3}$$

The aim of this chapter is to develop the theory—interesting in its own right—which leads to the proof of this theorem. The crucial step will be the theorem of Hartman–Grobman which implies that there are only $n$ distinct types of local behavior of a hyperbolic singularity.

In the analysis of the local behavior of the trajectories of $X$ at $p$ we will consider first the case where $p$ is a regular point, i.e., $X(p) \neq 0$, and then later the case where $p$ is singular point, i.e., $X(p) = 0$.

## 2.2 The flow box

In this section we show that any two differential equations at regular points are locally equivalent. This fact is an immediate consequence of the flow box theorem which states, in concise geometrical language, the classical theorems about existence, uniqueness and differentiability of solutions with respect to the initial data.

2.2.1 DEFINITION: A *flow box* of $X$ at $p$ is a pair $(F, y)$, where $F$ is neighborhood of $p$ in $D^n$ and $y : F \to R^n$ is a diffeomorphism mapping $F$ onto the unit cube $Q : |y_i| \leqslant 1$, $i = 1, \ldots, n$, and such that

$$(dy)(X) = (1, 0, \ldots, 0),$$

that is, the differential of $y$ takes the vector field $X$ in $F$ onto the unit field parallel to the $y_1$-axis.

2.2.2 FLOW BOX THEOREM: *At every regular point $p$ of a differential equation $X$ there is a flow box for $X$.*

We will not prove this theorem here. The proof is an easy consequence of the differentiability of solutions with respect to initial data and of the inverse function theorem; see [1].

Another analytical way to state the flow box theorem is this: *Given a regular point of a differential equation*

$$\dot{x}_i = dx_i/dt = X_i(x_1, \ldots, x_n), \quad i = 1, \ldots, n,$$

*there is a change of coordinates*

$$x_i = x_i(y_1, \ldots, y_n), \quad i = 1, \ldots, n$$

*defined on a neighborhood of that point which transforms the above differential equation into the equation*

$$dy_1/d\tau = 1, \quad dy_j/d\tau = 0, \quad j = 2, \ldots, n,$$

*where $\tau$ is another time parameter.*

Geometrically, the flow box theorem means that in the neighborhood of a regular point of a differential equation the integral curves can be "straightened out" by means of a change of coordinates. Because of this, it is sometimes said that near a regular point the trajectories of a differential equation are parallel or that we have a parallel flow. The first equation in (2.1.3) corresponds to this situation. Next we turn to the problem of classifying the local behavior of trajectories at singularities. As pointed out before, we will have to take the generic point of view.

## 2.3 The set $\mathcal{H}$ of equations with hyperbolic singularities

We now choose the set $\mathcal{H}$ needed in Theorem 2.1.2 to be the set of all differential equations in $\mathfrak{X}$ whose singularities are all hyper-

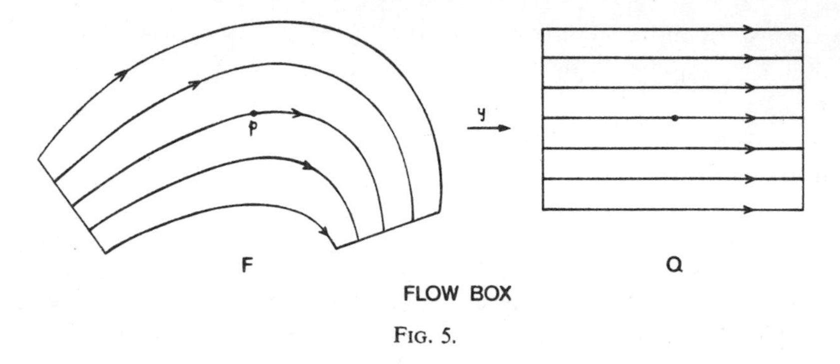


FIG. 5.

bolic. Recall that a singularity is hyperbolic if no eigenvalue of the Jacobian has zero real part (see section 1.3). The *index* of a singularity is the number of eigenvalues with negative real part.

We first remark that a differential equation $X = (X_1, \ldots, X_n) \in \mathcal{H}$ can only have finitely many singularities. Otherwise one singularity, say the origin, would be the limit of a sequence of singularities $p_i \to 0$, $i = 1, 2, \ldots$ . But since all singularities of $X$ are hyperbolic, the corresponding Jacobian matrices are non-singular, and so by the inverse function theorem, the vector field $(X_1, \ldots, X_n)$ defines a map $v : D^n \to R^n$ which is locally one-to-one. Then $v(0) = v(p_i) = 0$, $i = 1, 2, \ldots,$ contradicts the fact that $v$ is 1-1 at 0. A much stronger result of this type that can be obtained with a refinement of the inverse function argument is the following theorem. See [1, p. 305].

2.3.1 THEOREM: *If $p$ is a hyperbolic singularity of the differential equation $X$, and $W$ is any neighborhood of $p$ in $D^n$, then we can find a neighborhood $U \subset W$ of $p$ and a neighborhood $\mathfrak{U}$ of $X$ in $\mathfrak{X}$ such that, whenever $X' \in \mathfrak{U}$, then $X'$ has exactly one singularity, call it $p'$, in $U$. Moreover, $p'$ is hyperbolic and has the same index as $p$.*

Let us now prove that the differential equations in $\mathfrak{X}$ with hyperbolic singularities do indeed form an open, dense subset of $\mathfrak{X}$.

2.3.2 THEOREM: *The set $\mathcal{H}$ is open and dense in $\mathfrak{X}$.*

We first show that $\mathcal{H}$ is open. Let $X \in \mathcal{H}$ and let $p_1, \ldots, p_k$ be its singularities. From the previous theorem, we can find open disjoint neighborhoods $U_j$ of $\beta_j$, $j = 1, \ldots, k$, and corresponding neighborhoods $\mathfrak{U}_j$ of $X$ in $\mathfrak{X}$ such that whenever $X' \in \mathfrak{U}_j$, $X'$ has exactly one singularity, which is hyperbolic, in $U_j$. Now on the complement $C$ of the $U_j$'s, $X$ has no singularity. Since $C$ is compact we can thus find a neighborhood $\mathfrak{v}$ of $X$ such that whenever $X'' \in \mathfrak{v}$, then $X''$ has no singularity on $C$. Combining this fact with the previous one, we have that every $Y \in \mathfrak{v} \cap (\cap_{j=1}^{k} \mathfrak{U}_j) = \mathfrak{W}$ has exactly $k$ singularities, all hyperbolic, one inside each $U_j$. So the neighborhood $\mathfrak{W} \subset \mathcal{H}$ and $\mathcal{H}$ is open.

We now show that $\mathcal{H}$ is dense in $\mathfrak{X}$. For this we need some preliminaries. A singularity is said to be *simple* if no eigenvalue is zero or equivalently if its Jacobian matrix is nonsingular. Clearly, any hyperbolic singularity is also simple. The converse is naturally not true. The equation $\dot{x}_1 = x_2$, $\dot{x}_2 = -x_1$ has the origin as a singularity with eigenvalues $\pm i$, and so this singularity is simple but not hyperbolic. Such a singularity is called a *center*, since the nearby trajectories are concentric circles. One can show that simple singularities are isolated and so finite in number.

To prove that $\mathcal{H}$ is dense in $\mathfrak{X}$ we first show that the set $\mathcal{H}' \supset \mathcal{H}$ of differential equations with simple singularities is dense in $\mathfrak{X}$. Given any $X \in \mathfrak{X}$ and $\varepsilon > 0$, we need to show that there is $Z \in \mathcal{H}'$ with $d(X, Z) < \varepsilon$. From the Weierstrass approximation theorem, we can find a polynomial vector field $Y = (Y_1, \ldots, Y_n)$, i.e., the $Y_i$'s are polynomials, such that $d(X, Y) < \varepsilon/2$. The singularities of $Y$ may well not be simple; they may be common solutions to the polynomial equations

$$Y_1 = 0, \ldots, Y_n = 0, \quad \partial(Y_1, \ldots, Y_n)/\partial(x_1, \ldots, x_n) = 0. \tag{2.3.3}$$

We then find another polynomial differential equation $Z = (Z_1, \ldots, Z_n)$ with $d(Y, Z) < \varepsilon/2$ and such that the equation (2.3.3) corresponding to $Z$ has no common solution. Then $Z \in \mathcal{H}'$, and $d(X, Z) \leqslant d(X, Y) + d(Y, Z) < \varepsilon/2 + \varepsilon/2 = \varepsilon$. Since

$\varepsilon$ is arbitrary, this proves that $\mathcal{H}'$ is dense in $\mathfrak{X}$. Such a $Z$ is found by slightly changing the coefficients of $Y$ as follows. Let $a_1, \ldots, a_m$ be the totality of the coefficients of the polynomials $Y_1, \ldots, Y_n$. Then from a classical result of algebra, a necessary and sufficient condition in order that the $(n+1)$ polynomial equations (2.3.3) have a common solution $(x_1, \ldots, x_n)$ is that a certain polynomial relation, the resultant $F(a_1, \ldots, a_m) = 0$, be satisfied. It is easily seen that this relation is not identically satisfied (for example, by taking $Y_i = x_i$, $i = 1, \ldots, n$). But then the set of points $(a_1, \ldots, a_m) \in R^m$ for which $F(a_1, \ldots, a_m) \neq 0$ is open (an obvious fact) and dense in $R^m$. Density follows from an induction on $m$, the case $m = 1$ being the well-known fact that a polynomial of degree $d$ which has more than $d$ roots vanishes identically. So, by means of a slight change on the $a$'s we get a $Z$ with $d(Y, Z) < \varepsilon/2$ which does not satisfy (2.3.3).

Now we need to show that $\mathcal{H}$ is dense in $\mathcal{H}'$. Suppose we are given some $Z \in \mathcal{H}'$ and $\varepsilon > 0$. Because the singularities of $Z$ are isolated, we can make a small perturbation around each singularity, transforming it into a hyperbolic singularity, without introducing any new singularities. We construct such a perturbation as follows. Suppose that the origin is a simple singularity of $Z$. Then near 0 we have

$$\dot{x} = Z(x) = Ax + f(x), \quad x = (x_1, \ldots, x_n),$$

$$f(0) = 0, \quad df(0) = 0,$$

where $A$ is the non-singular Jacobian matrix at 0. Since the eigenvalues of $A + \lambda I$ are obtained by adding $\lambda$ to the eigenvalues of $A$, it is clear that for some $\lambda_0 > 0$, the matrix $A + \lambda I$, with $\lambda < \lambda_0$, has no eigenvalue with zero real part. It would follow that $\mathcal{H}$ is dense in $\mathcal{H}'$ if $Z$ were linear, i.e., $f = 0$. Since $f(x)/|x| \to 0$ when $|x| \to 0$, then we can find $b > a > 0$ such that inside the disc $D^n(b)$ of radius $b$ centered at the origin the only singularity of $Z$ is 0 and inside $D^n(a)$,

$$|Ax + f(x)| = |x| \left| A \frac{x}{|x|} + \frac{f(x)}{|x|} \right| \geqslant \alpha |x|,$$

where $\alpha$ is some positive number, say, half the norm $\|A\|$ of $A$. Then for $\lambda_1$ small enough, $\lambda < \lambda_1$ implies that

$$|(A + \lambda I)x + f(x)| \geqslant \frac{\alpha}{2}|x|, \tag{2.3.4}$$

so that inside $D^n(a)$, the left hand side of (2.3.4) vanishes only at the origin. Let

$$\beta = \sup_x |Ax + f(x)| > 0; \quad x \in D^n(b) - D^n(a).$$

Let $\phi$ be a differentiable function such that

$$\phi(x) = 1 \quad \text{if} \quad x \in D^n(a), \quad \phi(x) = 0 \quad \text{if} \quad x \notin D^n(b),$$

and consider the perturbation $W_0$ of $Z$, which vanishes outside $D^n(b)$, given by

$$\dot{x} = (A + \lambda\phi(x)I)x + f(x), \tag{2.3.5}$$

where $\lambda < \min(\lambda_0, \lambda_1)$ is so small that $d(Z, W_0) < \varepsilon$. For $x \in D^n(b) - D^n(a)$, the right hand side of (2.3.5) never vanishes and so it can only vanish in $D(a)$, i.e., at $x = 0$. Then $W_0$ has the origin as a hyperbolic singularity. Doing the same for the other singularities of $Z$ we get $W \in \mathcal{H}$ with $d(Z, W) < \varepsilon$. This proves Theorem 2.3.2.

## 2.4 The classification of hyperbolic singularities

We now return to Theorem 2.1.2 which solves the fundamental problem of the local generic theory. From the flow box theorem, all regular points of a differential equation are locally equivalent to a parallel flow, the first equation of 2.2.2. The proof of Theorem 2.1.2 then follows easily from our next theorem which classifies hyperbolic singularities.

2.4.1 THEOREM (Hartman–Grobman): *If $p$ is a hyperbolic singularity of index $s$, $0 \leqslant s \leqslant n$, of a differential equation $X \in \mathfrak{X}$, then locally this differential equation is equivalent to the "standard"* *hyperbolic singularity of index s, given by*

$$\dot{x}_1 = -x_1, \ldots, \dot{x}_s = -x_s; \dot{x}_{s+1} = x_{s+1}, \ldots, \dot{x}_n = x_n.$$

*This can also be written as*

$$\dot{\xi} = -\xi, \quad \dot{\eta} = \eta,$$

*where* $\xi = (\xi_1, \ldots, \xi_s)$, $\eta = (\eta_1, \ldots, \eta_u)$ *and* $x = (\xi, \eta)$.

This theorem was originally proved analytically [19, 20] in a slightly different form, saying that a hyperbolic singularity is locally equivalent to its linear part. For the geometric proof that will be given here, we need to make a digression about the concepts of "stable and unstable manifolds" of a hyperbolic singularity, a very important concept in its own right. Then we prove Theorem 2.4.1 in the case $s = n$, where the singularity is a sink; afterwards we consider the general case.

## 2.5 Stable and unstable manifolds of a hyperbolic singularity

Let the origin be a hyperbolic singularity of the differential equation $X \in \mathfrak{X}$, of index $s$. This means $s$ of the eigenvalues have negative real part and the other $n - s$ have positive real part.

Let $W^s$ be the set of all points $p \in D^n$ for which the trajectory through $p$ tends to the origin as $t \to \infty$; of course $0 \in W^s$. The *stable manifold theorem* says that $W^s$ is an *immersed submanifold of dimension s* of $D^n$. This means that there exists a map

$$f : R^s \to D^n, \quad f(R^s) = W^s,$$

which is one-to-one and of maximum rank, i.e., of rank $s$ everywhere; $f$ has the same class of differentiability as $X$. This $W^s$ is called the *stable manifold* of the hyperbolic singularity we are considering. For instance, in the 2-dimensional case considered in Figure 1b, where $s = 1$, the theorem says that there are exactly two trajectories entering the saddle (hyperbolic) point and that together with this singularity itself they form a differentiable curve. The existence of $W^s$ is proved first locally in the neighborhood of the origin, i.e., one shows the existence of a small piece of $W^s$ containing $f(0)$. This is done by analytical methods [2, p. 330]. Going backwards with respect to time along each trajectory of this small piece, we get the full $W^s$.

The set $W^s$ may be quite complicated. If we consider a small disc around a point of $W^s$ it may well be that the intersection of $W^s$ with this disc be composed of infinitely many connected components. In other words, $W^s$ may exhibit what is called *recurrence* in that it "comes back to itself" infinitely many times. However when, instead of all of $R^s$, we consider only a small neighborhood $U$ of a point $p \in R^s$, recurrence cannot occur. This means that the part of $W^s$ corresponding to $U$, i.e., $f(U)$, is such that given any point $q \in f(U)$, then a sufficiently small disc in $D^n$ centered at $q$ meets $f(U)$ in a single connected component. Such a map

$$f : U \subset R^s \to f(U) \subset D^n,$$

where $f$ is of maximum rank, is called a "regular surface" of dimension $s$ in $D^n$; it is also referred to as an "imbedded submanifold" of dimension $s$ in $D^n$.

In an entirely analogous way one defines the *unstable manifold* $W^u$ of our hyperbolic singularity: it is the set of all points $p$ in $D^n$ such that the trajectory of $X$ through $p$ tends to the singularity as $t \to -\infty$. This is the same as saying that $W^u$ is the stable manifold of the origin for the differential equation $-X$. Again $W^u$ is an immersed submanifold of dimension $u$ of $D^n$, i.e., there is a smooth one-to-one map

$$g : R^u \to D^n, \quad g(R^u) = W^u$$

of maximum rank $u$. Again $W^u$ may be quite complicated and may exhibit recurrence. When $s = 0$, $W^s$ is reduced to the singularity itself and $W^u$ is $n$-dimensional. Similarly, if $s = n$, $W^u = \{0\}$ and $W^s$ is $n$-dimensional.

From their very definitions, $W^s$ and $W^u$ meet only at the corresponding singularity and the stable (or unstable) manifolds of two distinct singularities never meet. But the stable manifold of one singularity $p$ may well intersect the unstable manifold of another singularity $q$, and even when $q = p$. In Figure 6 we indicate these possibilities. In case (a), $n = 3$, in case (b), $n = 2$.

We are now ready to prove Theorem 4.1 in the case $s = n$, when the hyperbolic singularity is a "sink"; the case $s = 0$, when it is a "source" is entirely analogous.

2.5.1 LEMMA: *Any sink is equivalent to the sink $X$ given by*

$$\dot{x}_i = -x_i, \quad i = 1, \ldots, n.$$

To prove this, let our sink $Y$ be

$$\dot{y} = Ay + f(y), \quad y = (y_1, \ldots, y_n), \quad f(0) = 0, \quad df(0) = 0, \tag{2.5.2}$$

where all the eigenvalues of $A$ have negative real part. From linear algebra we know [1, p. 145] that by choosing coordinates in (2.5.2),

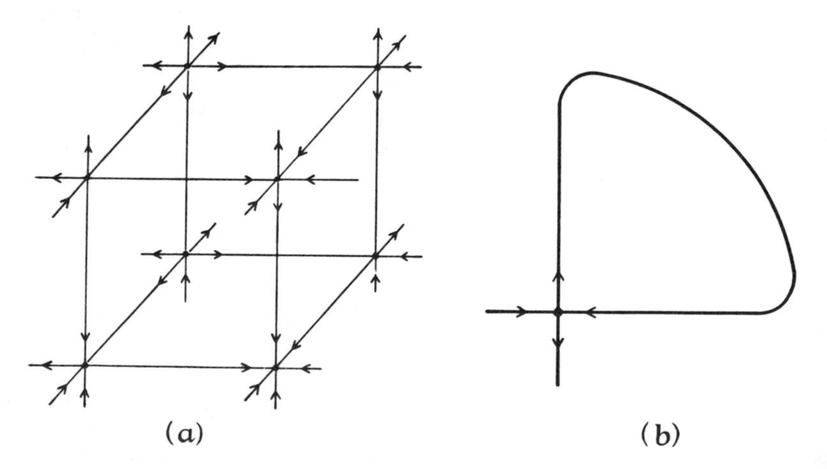


FIG. 6.

i.e., for a certain norm in Euclidean space $R^n$, the solutions of

$$\dot{y} = Ay$$

tend to zero monotonically, that is, they are transversal to the spheres centered at $y = 0$. From a simple compactness argument, one sees that for spheres of radii small enough the same property holds for solutions of $Y$. The proof of Lemma 2.5.1 is now an easy consequence of the following: Let $S_1$ be a sphere of radius $\rho_1$ centered at $x = 0$ and transversal to $X$, and let $S_2$ be a sphere of radius $\rho_2$ centered at $y = 0$ and transversal to $Y$. Call $D_1$ and $D_2$ the discs bounded by $S_1$ and $S_2$. We want to construct a homeomorphism

$$\phi : D_1 \to D_2,$$

mapping trajectories of $X$ onto trajectories of $Y$. From what we saw above, each point $x \in D_1$ is characterized by the pair $(x_0, r)$, where $r$ is the distance of $x$ from the origin and $x_0$ is the point where the trajectory of $X$ through $x$ meets $S_1$—going backwards in time. Points in $D_1$ have an analogous characterization. Now choose arbitrary homeomorphisms

$$h : S_1 \to S_2,$$

$$k : [0, \rho_1] \to [0, \rho_2].$$

The homeomorphism $\phi$ we want is defined by

$$\phi(x) = (h(x_0), k(r)).$$

If we represent a point $x \in D_1$ by $(x_0, t)$, where $x_0$ is as before and $t$ is the time spent along a trajectory of $X$ between $x_0$ and $x$, then we can define

$$\psi(x) = (h(x_0), t).$$

It is easy to verify that $\psi : D_1 \to D_2$ is continuous at the origin. Thus we get a homeomorphism that preserves both trajectories and time. This proves Lemma 2.5.1.

## 2.6 Proof of the Hartman–Grobman theorem

Hartman and Grobman, simultaneously and independently, gave essentially the same analytical proof of this theorem. The geometrical proof we sketch below is due to J. Palis [21].

Since we have already proved Theorem 2.4.1 for the case $s = 0$ and $s = n$, we need only to consider the case $0 < s < n$; $u = n - s$. Let $X$ represent the standard hyperbolic singularity of index $s$, which we will write as

$$\dot{\xi} = -\xi, \quad \dot{\eta} = \eta;$$

$$\xi = (\xi_1, \ldots, \xi_s), \quad \eta = (\eta_1, \ldots, \eta_u), \quad x = (\xi, \eta).$$

Let $W$ be another differential equation

$$\dot{W} = CW + h(W), \quad W = (W_1, \ldots, W_n)$$

exhibiting another hyperbolic singularity of index $s$ at the origin. This means that the matrix $C$ has $s$ eigenvalues with negative real part and $n - s$ eigenvalues with positive real part. Also

$$h(0) = 0, \quad dh(0) = 0.$$

Now from the proof of the existence of the stable and unstable manifolds of a hyperbolic singularity, it follows that by a suitable change of coordinates in the neighborhoods of $w = 0$, $w \to (y, z)$, the $y$-plane becomes the stable manifold whereas the $z$-plane becomes the unstable manifold. In other words, our equation $W$ can be written in coordinates $(y, z)$ as

$$\dot{y} = Ay + f(y, z)$$

$$\dot{z} = Bz + g(y, z)$$

with $y \in R^s$, $z \in R^u$, $s + u = n$, $A$ being an $s \times s$ matrix with all its eigenvalues with negative real part, and $B$ a $u \times u$ matrix with all its eigenvalues with positive real part. Further, $f$, $g$, and all their first-order partial derivatives vanish at the origin. The fact that the

$y$- and $z$-planes were chosen as the stable and unstable manifolds means that $f$ and $g$ must satisfy $f(y, 0) = 0$, $g(0, z) = 0$, respectively.

We want to construct a homeomorphism $H$ mapping a neighborhood of $w = (y, z) = 0$ onto a neighborhood of $x = (\xi, \eta) = 0$ and mapping trajectories of $W$ onto trajectories of $X$.

Proceeding as in Lemma 2.5.1 we consider spheres $S(y) = S^{s-1}(y)$ and $S(z) = S^{u-1}(z)$ situated, respectively, on the $y$- and $z$-planes and transversal to the vector field $W$. Here the upper index stands for the dimension of the spheres. Similarly let $\Sigma(\xi) = \Sigma^{s-1}(\xi)$ and $\Sigma(\eta) = \Sigma^{u-1}(\eta)$ be spheres on the $\xi$- and $\eta$-planes, respectively, and transversal to the vector field $X$. Let $\Delta(\xi)$ and $\Delta(\eta)$, $D(y)$ and $D(z)$ be discs whose boundaries are, respectively, $\Sigma(\xi)$ and $\Sigma(\eta)$, $S(y)$ and $S(z)$.

We now choose arbitrary homeomorphisms $l$ and $k$:

$$l : S(y) \to \Sigma(\xi), \quad k : S(z) \to \Sigma(\eta).$$

As shown in the preceding section, any point $p \in D(y)$ is determined by a certain point $y \in S(y)$ and by the time $t$ that the trajectory of $X$ takes to go from $y$ to $p$. We write $p = (y, t)$. Defining $l(p) = (l(x), p)$ and doing a similar thing with $k$, we see that the homeomorphisms $l$, $k$ between spheres extend to trajectory preserving homeomorphisms between the corresponding discs:

$$l : D(y) \to \Delta(\xi), \quad k : D(z) \to \Delta(\eta).$$

To get the desired trajectory preserving homeomorphism

$$H : (y, z) \to (\xi, \eta)$$

from a neighborhood of $(y, z) = 0$ onto a neighborhood of $(\xi, \eta) = 0$, we have to extend $l$ and $k$ beyond the stable and unstable manifolds.

To do this, we need a further argument. At every point $y$ of $S(y)$, let us put a $u$-dimensional disc $F_y^u$, normal to $S(y)$ and of radius $a$ so small that the $s - 1 + u = n - 1$ dimensional "fence" $S(y) \times F_y^u$ is transversal to the vector field $W$. This fence is said to be based on $S(y)$ with height $a$. The disc $F_y^u$ associated with $y \in S(y)$ is called the "fiber" over $y$. Similarly, we consider a fence $S(z) \times F_z^s$, with base $S(z)$, also transversal to the vector field $W$.

We now come to the crucial point of the proof, the so-called "λ-Lemma". If $t > 0$, let $W_t : D^n \to D^n$ be the map transforming each point $p \in D^n$ into the point $p(t) \in D^n$, obtained by moving $p$ along the trajectory of $W$ through $p$ for $t$ units of time. For each $t$, $W_t$ is a diffeomorphism [1]—this fact is a variant of the flow box theorem. We can define a similar map for $t < 0$, as long as $p(t) \in D^n$.

2.6.1 λ-LEMMA: *There is a neighborhood $V$ of $W = (y, z) = 0$ such that, given $\varepsilon > 0$, there exists $T > 0$ such that, for $t, t' > T$, $y \in S(y)$, $z \in S(z)$, the sets $W_t(F_y^u)$ and $W_{t'}(F_z^s)$ inside $V$ are $\varepsilon$-close in the $C^1$ sense to the planes $z$ and $y$, respectively.*

This means that inside $V$ not only is $W_t(F_y^u)$ $\varepsilon$-close to the $y$-plane but also its tangent plane makes always an angle less than $\varepsilon$ with the $z$-plane; similarly for $W_{t'}(F_z^s)$ and the $y$-plane. Although this lemma is fairly intuitive, its proof is complex and will not be given here.

From this lemma it follows that, for small $\varepsilon$, $W_t(F_y^u)$ and $W_{-t'}(F_z^s)$ inside $V$ behave like the $y$ and $z$-planes, i.e., they meet at a single point and can then be used as some kind of $y$ and $z$ coordinates for points in the neighborhood of $w = (y, z) = 0$. More precisely, every point $w$ in this neighborhood is characterized by $y \in S(y)$, $z \in S(z)$, $t$, $t'$ such that

$$w = W_t(F_y^u) \cap W_{-t'}(F_z^s) \cap V.$$

Of course, what we did in the neighborhood of $w = (y, z) = 0$ we can do in neighborhood of $x = (\xi, \eta) = 0$. So we define our desired homeomorphism $H$ by

$$H(w) = H(y, z) = W_t(F^u_{l(y)}) \cap W_{-t'}(F^s_{k(z)}).$$

The verification that this $H$ is a trajectory preserving homeomorphism is immediate.

## 3. GLOBAL THEORY

### 3.1 The 2-dimensional case: classification

In this last chapter we indicate briefly how the fundamental problem formulated in 1.6 can be solved in dimension 2. Then we discuss transversality and the Kupka–Smale theorem.

Theorem 1.4.2 states that the set $\Sigma$ of all differential equations in $D^2$, satisfying conditions (a), (b), (c) of Theorem 1.3.2, constitute an open dense subset of all differential equations. As mentioned in Section 1.6, this theorem solves the approximation problem for the generic theory in dimension 2. The proof of this assertion is part of the proof of the Kupka–Smale theorem which will be sketched below.

To say that Theorem 1.4.2 solves the approximation problem means that one should be able to classify all equivalence classes of $\Sigma$ modulo $\sim$. Let us indicate how this is done [10]. To simplify matters we will consider differential equations $X$ satisfying conditions (a), (c) of 1.3.2 and exhibiting no closed orbit. Such $X$ are said to be *gradient-like*. The case where there are closed orbits can be handled in a similar way. Call $\Sigma_0$ the set of all gradient-like differential equations in $\Sigma$. Our problem is then to classify $\Sigma_0$ modulo $\sim$.

We now associate with $X \in \Sigma_0$ a certain graph (network) $G(X)$, as follows. Let $\alpha_1, \ldots, \alpha_s$ be the sources of $X$, $\sigma_1, \ldots, \sigma_p$ its saddle points, and $\omega_1, \ldots, \omega_q$ its sinks. We are considering the boundary of $D^2$ as one of the sources. From index considerations [2, p. 399] one knows that $s, p, q$ must satisfy the index relation

$$s - p + q = 2.$$

Put now the $\alpha$'s, the $\sigma$'s and the $\omega$'s on three horizontal lines, in three levels. We now define $G(X)$ by giving a rule to put oriented edges between these points:

> If there is no saddle point, i.e., $p = 0$, then from the index relation, $s = q = 1$ and $G(X)$ is the oriented segment from $\alpha_1$ to $\omega_1$. If there are saddle points $\sigma$, then at each $\sigma$ draw two incoming edges to $\sigma$ from points $\alpha$ for which there is a trajectory leaving $\alpha$ and entering $\sigma$, and draw two outgoing edges from $\sigma$ to points $\omega$ for which there is a trajectory leaving $\sigma$ and entering $\omega$. See Figure 7, Figure 8, Figure 9. (3.1.1)

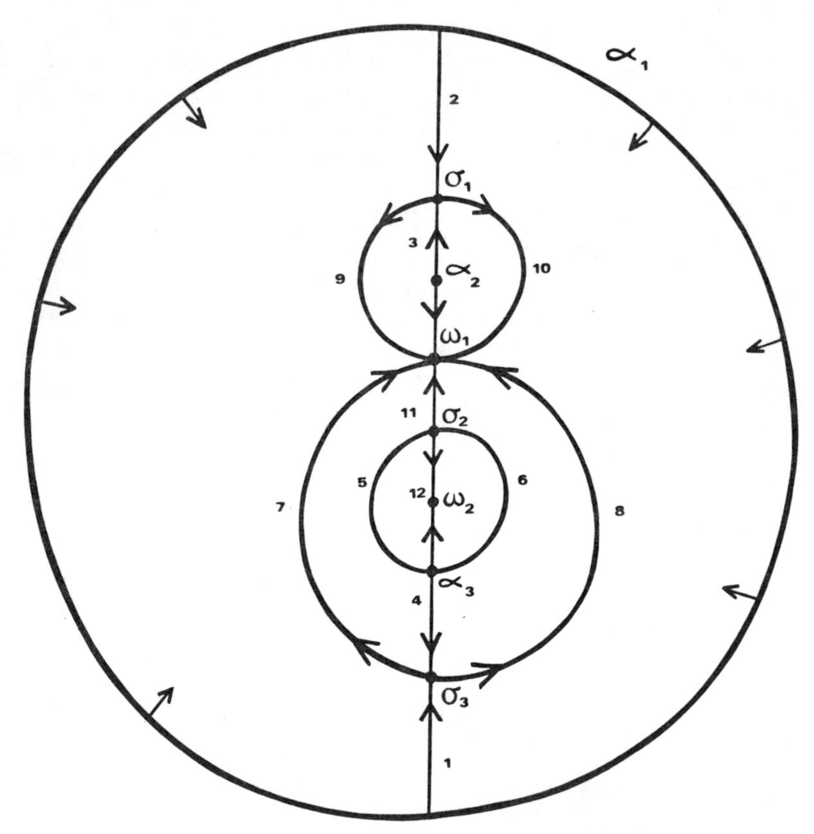


FIG. 7.

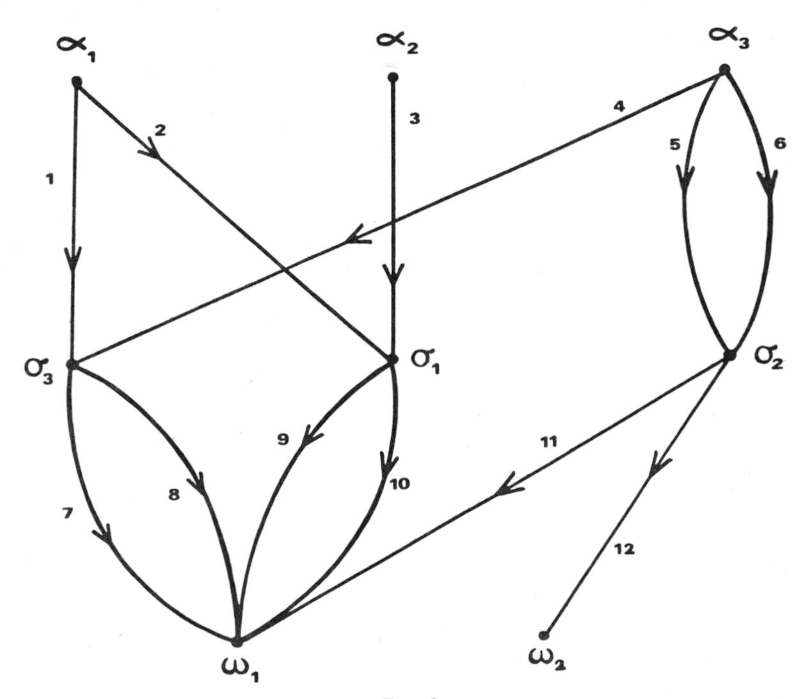
$\alpha_1$
$\alpha_2$
$\alpha_3$
1
2
3
4
5
6
$\sigma_3$
$\sigma_1$
$\sigma_2$
7
8
9
10
11
12
$\omega_1$
$\omega_2$

FIG. 8.

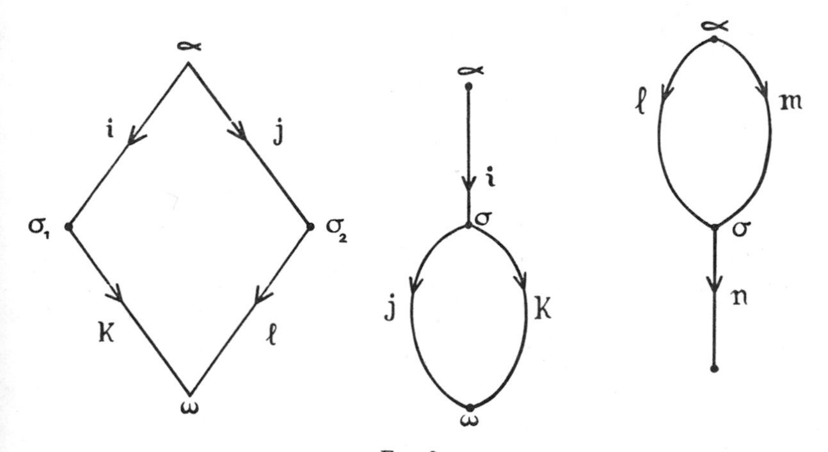
$\alpha$
i
j
$\sigma_1$
$\sigma_2$
K
ℓ
$\omega$
$\alpha$
i
$\sigma$
j
K
$\omega$
$\alpha$
ℓ
m
$\sigma$
n

FIG. 9.

For instance, if $X$ is the differential equation whose trajectories are represented in Figure 7, then $G = G(X)$ is the graph represented by Figure 8. We say that two such graphs, $G(X)$ and $G(Y)$ are isomorphic, $G(X) \cong G(Y)$ if there is a map between them transforming vertices onto vertices of the same level, edges onto edges, and preserving incidences.

There are examples showing that we may have $G(X) \cong G(Y)$ without having $X \sim Y$. We will next put the extra information on these graphs $G$ that will enable us to distinguish between the equivalence classes of $\Sigma_0$ modulo $\sim$.

Suppose that for a given $X \in \Sigma_0$, we delete from $D^2$ the sources, sinks and all stable and unstable manifolds associated with saddle points. The connected components of what is left correspond to the certain sets of edges of $G(X)$. For $X$ in Figure 7 and $G(X)$ in Figure 8, these sets represented in Figure 9 are called distinguished sets of type 1, 2 and 3, respectively. These distinguished sets are not arbitrarily situated in $G(X)$ but satisfy the following conditions:

(i) every edge of $G(X)$ belongs to exactly two distinguished sets, except the ones like $i$ in a distinguished set of type 2 or $n$ in a distinguished set to type 3;

(ii) if an edge like $i$ is incident to a certain $\alpha$, then no other edge of $G(X)$ is incident to this $\alpha$ and similarly for $\omega$ and $n$;

(iii) if two distinguished sets have in common a pair of edges, one entering and the other leaving the same saddle point, then they must coincide;

(iv) when more than one distinguished set is incident to a given vertex, then they can be written in cyclic order;

(v) the distinguished sets can be coherently oriented.

The graph $G(X)$, together with its distinguished sets satisfying (i)–(v), is called the *distinguished graph* of $X$ and denoted $G^*(X)$. Observe that there is a canonical way to associate $G^*(X)$ to $X$. We say that $G^*(X)$ and $G^*(Y)$ are isomorphic, $G^*(X) \cong G^*(Y)$, when there is an isomorphism between $G(X)$ and $G(Y)$ which preserves distinguished sets. In Figure 8, [3; 9, 10], [5, 8; 12], [4, 6; 8, 11], . . . , are distinguished sets.

From the proof in [23] that conditions (a), (b), (c) of Theorem

1.3.2 imply structural stability, it follows that:

*If* $X, Y \in \Sigma_0$, *then* $X \sim Y$ *if and only if* $G^*(X) \cong G^*(Y)$. (3.1.2)

Now given abstractly a graph $\Lambda$ with $s$ vertices $\alpha$, $p$ vertices $\sigma$ and $q$ vertices $\omega$, $s - p + q = 2$, satisfying (3.1.2) and conditions (i)–(v) above, we say that $G^*$ is an abstract distinguished graph. An abstract distinguished graph $\Lambda$ is said to be realizable in $D^2$ if we can find $X \in \Sigma_0$ such that $G^*(X) = \Lambda$.

3.1.3 THEOREM: *Every abstract distinguished graph* $\Lambda$ *is realizable in* $D^2$.

This theorem, together with (3.1.2), shows that to classify $\Sigma_0$ modulo $\sim$ it is enough to classify all abstract distinguished graphs. Clearly we can give a precise rule to exhibit all abstract distinguished graphs and consequenctly all equivalence classes of $\Sigma_0$ modulo $\sim$. The problem of computing the actual number of equivalence classes as a function of $s$, $p$, $q$ seems to be a difficult one.

## 3.2 Transversality

We now give a brief sketch of transversality so that we can state the theorem of Kupka–Smale. As in section 2.5, we let

$$f : R^p \to D^n, \quad p \leqslant n, \tag{3.2.1}$$

be one-to-one, of class $C^r$, and of maximum rank everywhere; $f(R^p) = U$ is what we called an immersed submanifold of $D^n$.

Similarly consider

$$g : R^q \to D^n, \quad q \leqslant n, \tag{3.2.2}$$

with $g(R^q) = V$ also an immersed submanifold of $D^n$ of dimension $q$.

3.2.3 DEFINITION: $U$ and $V$ are said to be *transversal at* $y \in U \cap V$ when the tangent spaces of $U$ and $V$ at $y$, together span the whole Euclidean space $R^n$,

$$T_y(U) + T_y(V) = R^n.$$

$U$ and $V$ are said to be *transversal* when either they are disjoint or else when they are transversal at every common point.

Recall that at a point $y \in U$, the *tangent space* $T_y(U)$ is defined as $df_x(R^p)$, where $f(x) = y$.

EXAMPLES:

(i) Two line segments in $D^2$ secant to each other are transversal;

(ii) two line segments secant to each other in $D^3$ are *not* transversal since their tangent spaces at the common point do span $R^2$ and not $R^3$;

(iii) two disjoint curves in $D^2$ are transversal, two curves in $D^2$ which are tangent at some point are not;

(iv) an open disc in $D^2$ and a curve contained in it are transversal.

From (iii) and (iv), we see that condition (c) of Theorem 1.3.2 implies that if $\sigma_1$ and $\sigma_2$ are any two singularities of $X$, then $W^s(\sigma_1)$ and $W^u(\sigma_2)$, the stable and unstable manifolds of $\sigma_1$ and $\sigma_2$, respectively, are transversal. This relation between singularities and transversality is exploited in the theorem of Kupka–Smale.

A key fact about transversality needed to prove the Kupka–Smale Theorem is Thom's transversality lemma. Roughly speaking, it says that if $U$ and $V$ are not transversal then, after perturbing $U$ slightly, they will become transversal. To make a more precise statement, we need to introduce the Whitney topology, on the space $\mathfrak{F}$ of all $C^r$-mappings $R^p \to D^n$. In this topology, a typical neighborhood of $f$ is the set of all $C^r$-maps $H : R^p \to D^n$ such that for some continuous function $\varepsilon : R^p \to R$, we have

$$d^r(f(x) - h(x)) < \varepsilon(x), \quad x \in R^p,$$

that is, at every $x \in R^p$, all components of $f - g$ and their partial derivatives of order $\leqslant r$ are less than $\varepsilon(x)$. With this topology, $\mathfrak{F}$ is not a metric space.

Referring to (3.2.1) and (3.2.2) we can now state the

3.2.4 TRANSVERSALITY LEMMA: *The subset of $\mathfrak{F}$ for which $f(R^p) = U$ is transversal to $V = g(R^q)$ is dense in $\mathfrak{F}$.*

## 3.3 The theorem of Kupka–Smale

We saw in Theorem 2.3.2 that the set $\mathcal{H}$ of differential equations in $D^n$ which have hyperbolic singularities is open and dense in the space $\mathfrak{X}$ of all differential equations, with the $C^r$-topology, $r \geqslant 1$. As remarked before, each $X \in \mathcal{H}$ has only a finite number of singularities and each has its own stable and unstable manifolds. And the intersections of these manifolds can be quite complicated.

We now focus attention on the closed orbits of $X$. The first thing that we have to do is to generalize to $n$ dimensions the concept, introduced for $n = 2$ in 1.3, of a hyperbolic closed orbit $\gamma$ of a differential equation $X$ in $D^n$. This generalization is straightforward. The cross-section $\sigma$ is now a small piece of a hyperplane, of dimension $n - 1$, and the Poincaré transformation $\pi : \sigma \rightarrow \sigma$ is defined the same way. *Hyperbolicity of* $\gamma$ means that no eigenvalue of $d\pi(0)$ has absolute value 1, where $0 \in \sigma$ is the point at which $\gamma$ meets $\sigma$.

As with hyperbolic singularities, we can associate with a hyperbolic closed orbit stable and unstable manifolds $W^s(\gamma)$ and $W^u(\gamma)$. $W^s(\gamma)$ is an immersed submanifold made up of all trajectories of $X$ which tend to $\gamma$ as $t \rightarrow \infty$ and $W^u(\gamma)$ is an immersed submanifold made up of all trajectories of $X$ which tend to $\gamma$ as $t \rightarrow -\infty$. These submanifolds have $\gamma$ as their intersection and they intersect transversally. The dimension of $W^s(\gamma)$ is 1 plus the number of eigenvalues of $d\pi(0)$ situated inside the unit circle. The dimension of $W^u(\gamma)$ is 1 plus the number of eigenvalues of $d\pi(0)$ situated outside the unit circle.

The arguments need to prove these facts are similar to those used with stable and unstable manifolds for a hyperbolic singularity.

At this point two questions naturally arise. Assuming $X$ has hyperbolic singularities, can we approximate $X$ by another differential equation $Y$ having hyperbolic closed orbits? Assuming this is possible, $Y$ then has one stable and one unstable manifold associated to each singularity and to each closed orbit. Can $Y$ be approximated by another $Z$ of the same type, i.e., its singularities and closed orbits are hyperbolic, such that all these stable and unstable manifolds are transversal? An affirmative answer to both questions is given by the following theorem:

3.3.1 THEOREM (Kupka–Smale [15, 16, 25]): *Let $\mathfrak{X}$ be the space of all differential equations in $D^n$ with the $C^r$-topology. Then there is a Baire set $\mathfrak{B}$ such that for $X \in \mathfrak{B}$, the singularities and closed orbits of $X$ are hyperbolic and the corresponding stable and unstable manifolds are transversal.*

It should be pointed out that for $X \in \mathfrak{B}$, the hyperbolic singularities are finite in number but there may well be an infinite number of hyperbolic closed orbits in $X$. Then we would have an infinite number of stable and unstable manifolds.

The basic approach for the proof of this theorem is as follows: Let $T > 0$ be an integer and call

$\mathfrak{X}(T)$ : the subset of $\mathcal{H}$ such that $X \in \mathfrak{X}(T)$ implies that all closed orbits of $X$ of period $\leqslant T$ are hyperbolic; and

$\tilde{\mathfrak{X}}(T)$ : the subset of $\mathfrak{X}(T)$ such that $X \in \tilde{\mathfrak{X}}(T)$ implies that the stable and unstable manifolds of all singularities and of all closed orbits of $X$ with period $\leqslant T$ are transversal.

Then $\mathfrak{B}$ will be defined as

$$\mathfrak{B} = \bigcap_{T=1}^{\infty} \tilde{\mathfrak{X}}(T).$$

Now to prove the theorem we have to prove the following two

steps:

(i) $\mathfrak{X}(T)$ is open and dense in $\mathfrak{X}$,

(ii) $\tilde{\mathfrak{X}}(T)$ is a Baire set in $\mathfrak{X}(T)$.

Both steps depend on transversality arguments. In the proof of (ii), at a crucial step we deform, via the transversality lemma, one stable manifold so as to be transversal to a fixed unstable manifold. But the transversality lemma has nothing to do with differential equations, and so one has to prove that this deformation of the stable manifold can be brought about by a perturbation of the differential equation.

## REFERENCES


1. M. Hirsch and S. Smale, *Differential Equations, Dynamical Systems and Linear Algebra*, Academic Press, New York, 1974.
2. E. A. Coddington and N. Levinson, *Theory of Ordinary Differential Equations*, McGraw-Hill, New York, 1955.
3. H. Poincaré, *Sur les courbes définies par une équation différentielle*, Oeuvres, vol. 1, Gauthier-Villars, Paris, 1928.
4. G. D. Birkhoff, *Collected Works*, 3 vols., Amer. Math. Soc., Providence.
5. A. M. Liapounov, *Problème général de la stabilité du mouvement*, Princeton University Press, 1947. (Reproduction of the French translation of 1907 of the Russian mémoire of 1892.)
6. H. Kneser, "Kurvensscharen auf den Ringflachen," *Math. Ann.*, **91** (1924), 135–154.
7. W. Kaplan, "Regular families filling the plane, II," *Duke Math. J.*, **8** (1941), 11–40.
8. A. Andronov and L. S. Pontrjagin, "Systèmes grossiers," *Dokl. Akad. Nauk SSSR*, **14** (1937), 247–251.
9. M. M. Peixoto, "On structural stability," *Ann. of Math.*, **69** (1959), 189–222.
10. ———, "On the classification of flows on 2-manifolds," in *Dynamical Systems* (M. M. Peixoto, ed.), Academic Press, New York, 1973, 389–419.
11. S. Smale, "Morse inequalities for a dynamical system," *Bull. Amer. Math. Soc.*, **66** (1960), 43–49.
12. D. Anosov, "Roughness of geodesic flows on compact Riemannian manifolds of negative curvature," *Sov. Math. Dokl.*, **3** (1962), 1068–1079.
13. S. Smale, "Structurally stable systems are not dense," *Amer. J. Math.*, **86** (1966), 491–496.

14. R. Williams, "The "DA" maps of Smale and structural stability," *Proc. Symp. Pure Math.*, vol. **XIV** (S. S. Chern and S. Smale ed.), 329–334, Amer. Math. Soc., Providence, 1970.

15. I. Kupka, "Contribution à la théorie des champs génériques," *Contrib. Diff. Eq.*, **2** (1963), 457–484.

16. S. Smale, "Stable manifolds for differential equations and diffeomorphisms," *Ann. Scuola Norm. Sup. Pisa*, **18** (1963), 97–116.

17. ———, "Differentiable dynamical systems," *Bull. Amer. Math. Soc.*, **73** (1967), 747–817.

18. R. Thom, *Stabilité structurelle et morphogénèse*, W. A. Benjamin, Reading, 1972.

19. P. Hartman, "A lemma in the structural stability of differential equations," *Proc. Amer. Math. Soc.*, **11** (1961).

20. D. Grobman, "Homeomorphisms of systems of differential equations," *Dokl. Akad. Nauk*, **128** (1965), 880–881.

21. J. Palis, "On Morse-Smale dynamical systems," *Topology*, **8** (1969), 385–404.

22. J. Palis and S. Smale, "Structural stability theorems," *Global Analysis, Proc. Symp. Pure Math.*, vol. XIV, Amer. Math. Soc., Providence, (1970).

23. M. C. Peixoto and M. M. Peixoto, "Structural stability in the plane with enlarged boundary conditions," *An. Acad. Brasil. Ci.*, **31** (1959), 135–160.

24. V. Guillemin and A. Pollack, *Differential Topology*, Prentice-Hall, Englewood Cliffs, 1974.

25. M. M. Peixoto, "On an approximation theorem of Kupka and Smale," *J. Differential Equations*, **3** (1967), 214–227.

# BOUNDARY VALUE PROBLEMS FOR ORDINARY DIFFERENTIAL EQUATIONS

*L. K. Jackson*

## 1. INTRODUCTION

We shall be concerned with ordinary differential equations in the normal form

$$y^{(n)} = f(x, y, y', y'', \ldots, y^{(n-1)}). \tag{1.1}$$

For such an equation the initial value problem deals with the existence of a solution $y(x)$ of (1.1) satisfying the conditions $y^{(i)}(x_0) = c_i$, $i = 0, 1, \ldots, n-1$, that is, the value of the solution and its successive derivatives are specified at one point. A problem which concerns the existence of a solution $y(x)$ of (1.1) which satisfies conditions placed upon $(y(x), y'(x), \ldots, y^{(n-1)}(x))$ at more than one point is customarily referred to as a boundary value problem. Obviously there are infinitely many types of boundary value problems that can be considered. For the most part we shall be concerned only with boundary value problems of the following type:

$$y^{(n)} = f(x, y, y', \ldots, y^{(n-1)}), \tag{1.1}$$

$$y^{(i)}(x_j) = c_{ji}, 0 \leqslant i \leqslant m_j - 1, 1 \leqslant j \leqslant k, \tag{1.2}$$

where $2 \leqslant k \leqslant n$, $x_1 < x_2 < \cdots < x_k$, $m_j \geqslant 1$ for each $1 \leqslant j \leqslant k$, and $\sum_{j=1}^{k} m_j = n$. We shall call such boundary value problems $k$-point boundary value problems.

The question of existence of solutions of initial value problems for (1.1) can be answered in a quite complete and concise manner [1]. Let $I$ be an interval of the real numbers and assume that $f(x, y, y', \ldots, y^{(n-1)})$ is continuous on the slab

$$S = I \times R^n.$$

Then, given any $x_0 \in I$ and any prescribed initial values at $x = x_0$, the corresponding initial value problem for (1.1) has a solution on some closed subinterval of $I$ containing $x_0$. If in addition, $f(x, y, y', \ldots, y^{(n-1)})$ satisfies a Lipschitz condition with respect to the variables $y, y', \ldots, y^{(n-1)}$ on each compact subset of $S$, then solutions of initial value problems are unique. Furthermore, if $f(x, y, y', \ldots, y^{(n-1)})$ satisfies a Lipschitz condition in the variables $y, y', \ldots, y^{(n-1)}$ on each closed sub-slab $[\alpha, \beta] \times R^n$ where $[\alpha, \beta]$ is a compact subinterval of $I$, then each initial value problem for (1.1) has a unique solution defined on the entire interval $I$.

On the other hand, the question of the existence of solutions of boundary value problems for equations of the form (1.1) is more complex. If $f(x, y, y', \ldots, y^{(n-1)})$ is assumed to be continuous on the slab $S = I \times R^n$, the Schauder fixed point theorem can be used to prove the existence of solutions of $k$-point boundary value problems provided suitable restrictions are placed on the points $x_1 < x_2 < \cdots < x_k$ and on the assigned boundary values as well. If in addition $f(x, y, y', \ldots, y^{(n-1)})$ satisfies a Lipschitz condition in the variables $y, y', \ldots, y^{(n-1)}$ on each sub-slab $[\alpha, \beta] \times R^n$ where $[\alpha, \beta]$ is a compact subinterval of $I$, then the Contraction Mapping Principle can be used to prove the existence of solutions of $k$-point problems with no restriction on the assigned boundary values provided $x_k - x_1$ is restricted to be sufficiently small. These "in the small" existence theorems will be discussed in Section 2.

When one attempts to establish the existence "in the large" of solutions of boundary value problems, that is, existence without restricting the interval length $x_k - x_1$, there are two main difficulties that arise. First, solutions of the differential equation may not extend to the interval $[x_1, x_k]$. For example, consider the boundary value problem

$$y'' = 1 + (y')^2,$$

$$y(0) = c_0, y(\pi) = c_1.$$

For this differential equation the initial value problem with initial values $y(0) = c_0$, $y'(0) = m$ has the unique solution

$$y = c_0 + \ln \sec(x + \operatorname{Arctan} m) - \ln \sec(\operatorname{Arctan} m).$$

Thus no solution of the differential equation extends to $[0, \pi]$ and the boundary value problem has no solution. To avoid this difficulty it isn't necessary that all solutions extend to $[x_1, x_k]$, merely that one that fits the boundary conditions does. In Section 3 we shall consider some methods by means of which this can be achieved. In particular, we shall impose conditions on the differential equation that imply that solutions either do extend or, if not, become unbounded. It can then be shown that a boundary value problem will have a solution if there exist certain types of solutions of two associated differential inequalities.

The second main type of difficulty in solving boundary value problems in the large is that, even though all solutions of the differential equation may extend, there may be gaps in the boundary values that are assumed. For example, let (1.1), (1.2) be a given $k$-point boundary value problem and let $Y$ be the set of all solutions of (1.1) that satisfy the $n - 1$ boundary conditions

$$y^{(i)}(x_j) = c_{ji}, 0 \leqslant i \leqslant m_j - 1, 1 \leqslant j \leqslant k - 1,$$

$$y^{(i)}(x_k) = c_{ki}, 0 \leqslant i \leqslant m_k - 2.$$

Then, even if all solutions of (1.1) extend and $Y$ is not empty, it may be that $\{y^{(m_k-1)}(x_k)|y \in Y\}$ is not the set of all real numbers so that for some values of $c_{k,m_k-1}$ the boundary value problem does not have a solution.

For linear equations it is quite easy to show that this type of behavior depends on the uniqueness of the solutions of the particular $k$-point problem under consideration. Consider the linear differential equation

$$y^{(n)} = \sum_{i=0}^{n-1} p_i(x)y^{(i)} + q(x), \tag{1.3}$$

where $q(x)$ and $p_i(x)$, $0 \leqslant i \leqslant n-1$, are continuous on $[a, b]$. With $a \leqslant x_1 < x_2 < \cdots < x_k \leqslant b$ the boundary conditions (1.2) are special cases of boundary conditions of the form

$$\phi_i(y) = c_i, \; 1 \leqslant i \leqslant n, \tag{1.4}$$

where each $\phi_i$ is a linear functional on $C^{(n-1)}[a, b]$. Now assume we are dealing with a particular boundary value problem (1.3), (1.4). Let $y_j(x)$, $1 \leqslant j \leqslant n$, be linearly independent solutions of the homogeneous equation

$$y^{(n)} = \sum_{i=0}^{n-1} p_i(x)y^{(i)} \tag{1.5}$$

and let $z(x)$ be a particular solution of the nonhomogeneous equation (1.3). Then every solution of (1.3) is of the form

$$y(x) = \sum_{j=1}^{n} \alpha_j y_j(x) + z(x),$$

where the $\alpha_j$, $1 \leqslant j \leqslant n$, are constants. Hence, the boundary value problem (1.3), (1.4) has a solution if and only if the constants $\alpha_j$ can be chosen to satisfy the system of linear equations

$$\sum_{j=1}^{n} \alpha_j \phi_i(y_j) = c_i - \phi_i(z), \; 1 \leqslant i \leqslant n. \tag{1.6}$$

Thus we conclude that the problem (1.3), (1.4) has a solution for every possible choice of the boundary values $c_i$, $1 \leqslant i \leqslant n$, if and only if the coefficient determinant, $\det(\phi_i(y_j))$, is different from zero. This is the case if and only if the homogeneous problem

$$y^{(n)} = \sum_{i=0}^{n-1} p_i(x) y^{(i)},$$

$$\phi_i(y) = 0, \; 1 \leqslant i \leqslant n,$$

has only the trivial solution $y(x) \equiv 0$. Finally, this in turn is the case if and only if solutions of the boundary value problem (1.3), (1.4), when they exist, are unique.

It follows that for a linear equation (1.3) and for a particular $k$-point problem, that is, for a particular choice of the points $x_1 < x_2 < \cdots < x_k$ and the integers $m_j$, $1 \leqslant j \leqslant k$, the uniqueness of solutions of the problem implies the existence of solutions. For example, the 2-point problem

$$y'' = p_0(x)y + p_1(x)y' + q(x),$$

$$y(x_1) = c_1, y(x_2) = c_2,$$

has a solution for every choice of $c_1$ and $c_2$ if and only if $u(x_2) \neq 0$, where $u(x)$ is the solution of the initial value problem

$$y'' = p_0(x)y + p_1(x)y',$$

$$y(x_1) = 0, y'(x_1) = 1.$$

The argument that we have given, that for linear differential equations the uniqueness of solutions of a $k$-point problem implies the existence of solutions, is an algebraic argument based on the linear structure of the solution set and on the linearity of the boundary conditions. Consequently, one might not expect "uniqueness implies existence" results to be valid for nonlinear differential equations. Indeed, it does seem to be too much to expect for this to be the case for any one fixed $k$-point boundary value problem for a nonlinear differential equation. However, it can be shown that, with suitable conditions imposed on a nonlinear $n$th order differential equation (1.1), the uniqueness of solutions of all $n$-point boundary value problems on an interval implies the existence of solutions of all $n$-point problems on that interval. The conditions include an assumption of extendability of solutions of the differential equation and a compactness condition on bounded sets of solutions. These results will be discussed in Section 4.

## 2. APPLICATIONS OF FIXED POINT THEOREMS

In order to apply a fixed point theorem to establish the existence of a solution of a boundary value problem, we must determine a suitable mapping of a function space into itself which is such that a fixed point of the mapping is a solution of the boundary value problem. We will consider first the case in which equation (1.1) is a second order differential equation.

If $h(x)$ is continuous on $[x_1, x_2]$, all solutions of

$$y'' = h(x)$$

are of the form

$$y = c_0 + c_1 x + \int_{x_1}^{x} (x - s)h(s)\, ds. \tag{2.1}$$

Consequently, the unique solution of the boundary value problem

$$y'' = h(x), \tag{2.2}$$

$$y(x_1) = 0, y(x_2) = 0, \tag{2.3}$$

can be written in the form (2.1) with appropriate values for the constants $c_0$, $c_1$. When these values of $c_0$ and $c_1$ are determined it is easily shown that the solution $y(x)$ of (2.2), (2.3) can be written in the form

$$y(x) = \int_{x_1}^{x_2} G(x, s)h(s)\, ds,$$

where the function $G(x, s)$, which is called the Green's function of the boundary value problem $y'' = 0, y(x_1) = 0, y(x_2) = 0$, is defined by

$$G(x, s) = \begin{cases} \dfrac{(x - x_1)(s - x_2)}{x_2 - x_1}, & \text{for } x_1 \leqslant x \leqslant s, \\ \dfrac{(x - x_2)(s - x_1)}{x_2 - x_1}, & \text{for } s \leqslant x \leqslant x_2. \end{cases}$$

Likewise the solution of the boundary value problem

$$y'' = h(x),$$
$$y(x_1) = y_1, y(x_2) = y_2,$$

can be written in the form

$$y(x) = \int_{x_1}^{x_2} G(x, s)h(s)\, ds + \omega(x),$$

where $\omega(x)$ is the solution of $y'' = 0$ satisfying $\omega(x_1) = y_1$ and $\omega(x_2) = y_2$.

Now suppose that $f(x, y, y')$ is continuous on $[x_1, x_2] \times R^2$. Then from the above representation of solutions it is clear that, if $\omega(x)$ is the solution of $y'' = 0$ with $\omega(x_1) = y_1$ and $\omega(x_2) = y_2$, then a function $y(x)$ belongs to $C^{(2)}[x_1, x_2]$ and is a solution of the boundary value problem

$$y'' = f(x, y, y'), \tag{2.4}$$

$$y(x_1) = y_1, y(x_2) = y_2, \tag{2.5}$$

if and only if $y(x)$ belongs to $C^{(1)}[x_1, x_2]$ and is a solution of the integral equation

$$y(x) = \int_{x_1}^{x_2} G(x, s) f(s, y(s), y'(s))\, ds + \omega(x) \tag{2.6}$$

on $[x_1, x_2]$. Thus, if $T : C^{(1)}[x_1, x_2] \to C^{(1)}[x_1, x_2]$ is defined by

$$(Ty)(x) = \int_{x_1}^{x_2} G(x, s) f(s, \dot{y}(s), y'(s))\, ds + \omega(x)$$

for all $y \in C^{(1)}[x_1, x_2]$ and $x_1 \leqslant x \leqslant x_2$, then a fixed point of $T$ is a solution of (2.6), hence, also of the boundary value problem (2.4), (2.5).

We consider first the case where $f(x, y, y')$ is continuous on $[\alpha, \beta] \times R^2$, $[\alpha, \beta]$ is a compact interval, and $f(x, y, y')$ satisfies a Lipschitz condition in the variables $y, y'$ on $[\alpha, \beta] \times R^2$. In this case we will show that all boundary value problems (2.4), (2.5) have unique solutions provided $\alpha \leqslant x_1 < x_2 \leqslant \beta$ and $x_2 - x_1$ is sufficiently small. The proof of this result makes use of the Contraction Mapping Principle [2, p. 14].

THE CONTRACTION MAPPING PRINCIPLE: *Assume that $\langle X, d \rangle$ is a complete metric space and that $T : X \to X$ is a mapping with the property that there is a constant $k$ such that $0 \leqslant k < 1$ and such that*

$$d(Ty, Tz) \leqslant k d(y, z)$$

*for all $y, z \in X$. Then $T$ has a unique fixed point, that is, there is a unique $y_0 \in X$ such that $Ty_0 = y_0$.*

THEOREM 2.1: *Assume that $f(x, y, y')$ is continuous on $[a, b] \times R^2$ and satisfies the Lipschitz condition*

$$|f(x, y_1, y_2) - f(x, z_1, z_2)| \leqslant P_0|y_1 - z_1| + P_1|y_2 - z_2|$$

*for all $(x, y_1, y_2)$, $(x, z_1, z_2) \in [a, b] \times R^2$, where $P_0$ and $P_1$ are fixed positive numbers. Then for any $y_1$ and $y_2$ the boundary value problem* (2.4), (2.5) *has a unique solution provided $a \leqslant x_1 < x_2 \leqslant b$ and $x_2 - x_1 < \delta$ where $\delta$ is the positive solution of*

$$\tfrac{1}{8} P_0 \delta^2 + \tfrac{1}{2} P_1 \delta = 1. \tag{2.7}$$

*Proof:* For any $a \leqslant x_1 < x_2 \leqslant b$ the space $C^{(1)}[x_1, x_2]$ is a Banach space with respect to the norm defined by

$$\|y\| = P_0\|y\|_\infty + P_1\|y'\|_\infty$$

where, for $g \in C[x_1, x_2]$, $\|g\|_\infty = \text{Max}\{|g(x)| \,|\, x_1 \leqslant x \leqslant x_2\}$. Now we assume that we are dealing with a fixed boundary value problem (2.4), (2.5) with corresponding integral equation (2.6) and we define the mapping $T : C^{(1)}[x_1, x_2] \to C^{(1)}[x_1, x_2]$ by

$$(Ty)(x) = \int_{x_1}^{x_2} G(x, s) f(s, y(s), y'(s))\, ds + \omega(x)$$

for each $y \in C^{(1)}[x_1, x_2]$ and all $x_1 \leqslant x \leqslant x_2$. Then for all $y, z \in C^{(1)}[x_1, x_2]$ and all $x_1 \leqslant x \leqslant x_2$ we have

$$\begin{aligned}
&|(Ty)(x) - (Tz)(x)| \\
&\qquad \leqslant \int_{x_1}^{x_2} |G(x, s)|\, |f(s, y(s), y'(s)) - f(s, z(s), z'(s))|\, ds \\
&\qquad \leqslant \int_{x_1}^{x_2} |G(x, s)|\, ds (P_0\|y - z\|_\infty + P_1\|y' - z'\|_\infty).
\end{aligned}$$

Similarly, for all $x_1 \leqslant x \leqslant x_2$

$$\begin{aligned}
&|(Ty)'(x) - (Tz)'(x)| \\
&\qquad \leqslant \int_{x_1}^{x_2} \left| \frac{\partial}{\partial x} G(x, s) \right| ds (P_0\|y - z\|_\infty + P_1\|y' - z'\|_\infty).
\end{aligned}$$

By an elementary calculation it can be established that

$$\underset{x_1 \leqslant x \leqslant x_2}{\text{Max}} \int_{x_1}^{x_2} |G(x, s)|\, ds = \tfrac{1}{8}(x_2 - x_1)^2 \tag{2.8}$$

and

$$\underset{x_1 \leqslant x \leqslant x_2}{\text{Max}} \int_{x_1}^{x_2} \left| \frac{\partial}{\partial x} G(x, s) \right| ds = \tfrac{1}{2}(x_2 - x_1). \tag{2.9}$$

Hence

$$\|Ty - Tz\|_\infty \leqslant \tfrac{1}{8}(x_2 - x_1)^2(P_0\|y - z\|_\infty + P_1\|y' - z'\|_\infty)$$

and

$$\|(Ty)' - (Tz)'\|_\infty \leqslant \tfrac{1}{2}(x_2 - x_1)(P_0\|y - z\|_\infty + P_1\|y' - z'\|_\infty)$$

from which it follows that

$$\|Ty - Tz\| \leqslant \left(\tfrac{1}{8}P_0(x_2 - x_1)^2 + \tfrac{1}{2}P_1(x_2 - x_1)\right)\|y - z\|.$$

It follows that, if $x_2 - x_1 < \delta$ where $\delta$ is the positive root of (2.7), then $T$ is a contraction mapping and has a unique fixed point in $C^{(1)}[x_1, x_2]$.

In the next existence theorem, again for the 2-point boundary value problem for a second order differential equation, we assume only the continuity of the function $f(x, y, y')$. In this case we obtain the existence of solutions on sufficiently small intervals provided the boundary values are restricted as well. The proof requires a form of the more powerful Schauder fixed point theorem [2, p. 96].

SCHAUDER THEOREM: *If $K$ is a closed convex subset of a Banach space $B$, if $T : K \to K$ is continuous on $K$, and if $\overline{T(K)}$ is a compact subset of $B$, then $T$ has a fixed point in $K$.*

THEOREM 2.2: *Assume that $f(x, y, y')$ is continous on $[a, b] \times R^2$. Then given $M > 0$ and $N > 0$ there is a $\delta = \delta(M, N) > 0$ such that the boundary value problem* (2.4), (2.5) *has a solution provided $a \leqslant x_1 < x_2 \leqslant b$, $x_2 - x_1 \leqslant \delta$, $|y_1| \leqslant M$, $|y_2| \leqslant M$, and $|y_1 - y_2| \leqslant N|x_1 - x_2|$.*

*Proof:* As in the proof of Theorem 2.1, it suffices to show that the mapping $T : C^{(1)}[x_1, x_2] \to C^{(1)}[x_1, x_2]$ defined by

$$(Ty)(x) = \int_{x_1}^{x_2} G(x, s) f(s, y(s), y'(s))\, ds + \omega(x)$$

has a fixed point. The space $C^{(1)}[x_1, x_2]$ is also a Banach space with respect to the norm

$$\|y\| = \|y\|_\infty + \|y'\|_\infty$$

which is more convenient in the proof of this Theorem. Let

$$Q = \operatorname{Max}\{|f(x, y, y')| \,|\, a \leqslant x \leqslant b, |y| \leqslant 2M, |y'| \leqslant 2N\}. \tag{2.10}$$

Assume that $a \leqslant x_1 < x_2 \leqslant b$ and that the boundary values $y_1, y_2$ satisfy $|y_1| \leqslant M$, $|y_2| \leqslant M$, and $|y_1 - y_2| \leqslant N|x_1 - x_2|$. Then it follows that the linear function $\omega(x)$ satisfies $|\omega(x)| \leqslant M$ and $|\omega'(x)| \leqslant N$ on $[x_1, x_2]$.

Let $K$ be the subset of $C^{(1)}[x_1, x_2]$ defined by

$$K = \{y \in C^{(1)}[x_1, x_2] \,|\, |y(x)| \leqslant 2M, |y'(x)| \leqslant 2N \text{ on} [x_1, x_2]\}.$$

Then it is clear that $K$ is a closed bounded convex subset of $C^{(1)}[x_1, x_2]$. For any $y \in K$ and for all $x_1 \leqslant x \leqslant x_2$ we have

$$\begin{aligned} |(Ty)(x)| &\leqslant \int_{x_1}^{x_2} |G(x, s)|\, |f(s, y(s), y'(s))|\, ds + |\omega(x)| \\ &\leqslant Q \int_{x_1}^{x_2} |G(x, s)|\, ds + M \end{aligned}$$

and

$$\begin{aligned} |(Ty)'(x)| &\leqslant \int_{x_1}^{x_2} \left| \frac{\partial}{\partial x} G(x, s) \right| |f(s, y(s), y'(s))|\, ds + |\omega'(x)| \\ &\leqslant Q \int_{x_1}^{x_2} \left| \frac{\partial}{\partial x} G(x, s) \right| ds + N. \end{aligned}$$

Thus applying (2.8) and (2.9) we conclude that $T$ maps $K$ into $K$ if $\frac{1}{8} Q(x_2 - x_1)^2 \leqslant M$ and $\frac{1}{2} Q(x_2 - x_1) \leqslant N$, that is, if $(x_2 - x_1)$

$\leqslant \delta(M, N)$, where

$$\delta(M, N) = \text{Min}\left\{\sqrt{\frac{8M}{Q}}, \frac{2N}{Q}\right\}. \tag{2.11}$$

The continuity of $T$ on $K$ follows from the fact that $f(x, y, y')$ is uniformly continuous on the set which is used in defining $Q$ in (2.10). It follows from the definition of $T$ that for any $y \in K$ and any $x_1 \leqslant x \leqslant x_2$

$$|(Ty)''(x)| = |f(x, y(x), y'(x))| \leqslant Q.$$

Hence, if $x_2 - x_1 \leqslant \delta(M, N)$, the sets $\{(Ty)'|y \in K\}$ and $\{Ty|y \in K\}$ are uniformly bounded and equicontinuous on $[x_1, x_2]$. The fact that $\overline{T(K)}$ is compact then follows from the Ascoli-Arzela Theorem. The Schauder Theorem is applicable and a fixed point of $T$ is a solution of the boundary value problem. The solution is not necessarily unique.

In Section 3 we will be concerned with the important special case in which $f(x, y, y')$ is bounded on $[a, b] \times R^2$. The following corollary of Theorem 2.2 establishes the global existence of solutions of 2-point boundary value problems in this case.

COROLLARY 2.3: *If $f(x, y, y')$ is bounded and continuous on $[a, b] \times R^2$, then, for any $a \leqslant x_1 < x_2 \leqslant b$ and any $y_1, y_2$, the boundary value problem* (2.4), (2.5) *has a solution.*

*Proof:* Let

$$Q = \sup\{|f(x, y, y')| \,|\, (x, y, y') \in [a, b] \times R^2\}.$$

Assume that $x_1$, $x_2$ are given with $a \leqslant x_1 < x_2 \leqslant b$ and that $y_1, y_2$ in the boundary values $y(x_1) = y_1, y(x_2) = y_2$ are specified. Then, if $M > 0$ and $N > 0$ are chosen large enough to satisfy the conditions $|y_1| \leqslant M$, $|y_2| \leqslant M$, $\frac{1}{8}Q(x_2 - x_1)^2 \leqslant M$, $|y_1 - y_2| \leqslant N|x_1 - x_2|$, and $\frac{1}{2}Q(x_2 - x_1) \leqslant N$, it follows that $x_2 - x_1 \leqslant \delta(M, N)$ and the existence of a solution of the boundary value problem follows from Theorem 2.2.

Next we see how arguments similar to those used above can be employed to prove local existence theorems for $k$-point boundary value problems for differential equations of arbitrary order. First, it is easy to see that no nontrivial solution of $y^{(n)} = 0$ has $n$ zeros, counting multiplicities of zeros. Hence, the $k$-point homogeneous boundary value problem

$$y^{(n)} = 0, \tag{2.12}$$

$$y^{(i)}(x_j) = 0, 0 \leqslant i \leqslant m_j - 1, 1 \leqslant j \leqslant k, \tag{2.13}$$

has only the trivial solution $y(x) \equiv 0$. It follows, see for example [3, p. 190–193], that the problem (2.12), (2.13) has a Green's function $G(x, s)$ which is characterized by the following conditions:

(i) $\dfrac{\partial^r}{\partial x^r} G(x, s)$ is continuous on $[x_1, x_k] \times [x_1, x_k]$ for $r = 0, 1, \ldots, n-2$ and $\dfrac{\partial^{n-1}}{\partial x^{n-1}} G(x, s)$ is continuous on each of the closed triangles $x_1 \leqslant x \leqslant s \leqslant x_k$ and $x_1 \leqslant s \leqslant x \leqslant x_k$.

(ii) $\dfrac{\partial^{n-1}}{\partial x^{n-1}} G(s+0, s) - \dfrac{\partial^{n-1}}{\partial x^{n-1}} G(s-0, s) = 1$ for $x_1 < s < x_k$.

(iii) For each fixed $s$, $G(x, s)$ satisfies $y^{(n)} = 0$ as a function of $x$ for $x \neq s$.

(iv) For each fixed $s$, $G(x, s)$ as a function of $x$ satisfies the boundary conditions (2.13).

Furthermore, if $h(x)$ is continuous on $[x_1, x_k]$, the unique solution of the boundary value problem

$$y^{(n)} = h(x), \tag{2.14}$$

$$y^{(i)}(x_j) = c_{ji}, 0 \leqslant i \leqslant m_j - 1, 1 \leqslant j \leqslant k, \tag{2.15}$$

is given by

$$y(x) = \int_{x_1}^{x_2} G(x, s) h(s)\, ds + \omega(x),$$

where now $\omega(x)$ is the polynomial of degree $n-1$ that is the solution of $y^{(n)} = 0$ which satisfies the boundary conditions $\omega^{(i)}(x_j) = c_{ji}$, $0 \leqslant i \leqslant m_j - 1$, $1 \leqslant j \leqslant k$.

The fact that the properties (i)–(iv) characterize the Green's function can be used to obtain a rather simple expression for the Green's function for the problem (2.12), (2.13). First let $u_r(x)$, $1 \leqslant r \leqslant n$, be solutions of $y^{(n)} = 0$ such that for $1 \leqslant r \leqslant m_1$, $u_r(x)$ satisfies the boundary conditions

$$u_r^{(i)}(x_1) = 0, 0 \leqslant i \leqslant m_1 - 1, i \neq r - 1,$$

$$u_r^{(r-1)}(x_1) = 1,$$

$$u_r^{(i)}(x_j) = 0, 0 \leqslant i \leqslant m_j - 1, 2 \leqslant j \leqslant k,$$

and for $\Sigma_{j=1}^{l-1} m_j < r \leqslant \Sigma_{j=1}^{l} m_j$, $2 \leqslant l \leqslant k$, $p = r - \Sigma_{j=1}^{l-1} m_j$, $u_r(x)$ satisfies the boundary conditions

$$u_r^{(i)}(x_l) = 0, 0 \leqslant i \leqslant m_l - 1, i \neq p - 1,$$

$$u_r^{(p-1)}(x_l) = 1,$$

$$u_r^{(i)}(x_j) = 0, 0 \leqslant i \leqslant m_j - 1, 1 \leqslant j \leqslant k, j \neq l.$$

Thus the functions $u_r(x)$ are solutions of $y^{(n)} = 0$ satisfying the boundary conditions (2.15) in which one of the $c_{ji}$'s has the value 1, the other $c_{ji}$'s have the value 0, and the numbering $1 \leqslant r \leqslant n$ corresponds to the position of the boundary value 1 in a consecutive numbering of the boundary conditions (2.15). For any fixed $s$, $\dfrac{(x-s)^{n-1}}{(n-1)!}$ is a solution of $y^{(n)} = 0$ as a function of $x$. For $1 \leqslant r \leqslant m_1$ let

$$v_r(s) = \frac{(x_1 - s)^{n-r}}{(n-r)!}$$

and for $\Sigma_{j=1}^{l-1} m_j < r \leqslant \Sigma_{j=1}^{l} m_j$, $2 \leqslant l \leqslant k$, $p = r - \Sigma_{j=1}^{l-1} m_j$, let

$$v_r(s) = \frac{(x_l - s)^{n-p}}{(n-p)!}.$$

Then

$$\sum_{r=1}^{n} v_r(s)u_r(x)$$

as a function of $x$ is a solution of $y^{(n)} = 0$ and has the same (2.15) boundary values as $\dfrac{(x-s)^{n-1}}{(n-1)!}$. Since solutions of (2.12), (2.15) are unique, it follows that

$$\frac{(x-s)^{n-1}}{(n-1)!} \equiv \sum_{r=1}^{n} v_r(s)u_r(x)$$

with this an identity in both $x$ and $s$. Now for $1 \leqslant j < k$, $x_j \leqslant s \leqslant x_{j+1}$, and $\rho_j = \Sigma_{t=1}^{j} m_t$, define $G(x, s)$ by

$$G(x,s) = \begin{cases} -\displaystyle\sum_{r=\rho_j+1}^{r=n} v_r(s)u_r(x), & \text{for } x_1 \leqslant x \leqslant s, \\ +\displaystyle\sum_{r=1}^{r=\rho_j} v_r(s)u_r(x), & \text{for } s \leqslant x \leqslant x_k. \end{cases}$$

With $G(x, s)$ so defined it is easy to verify that the conditions (i)–(iv) are satisfied, hence, this $G(x, s)$ is the Green's function for the problem (2.12), (2.13).

As a consequence of the remarks made concerning the boundary value problem (2.14), (2.15) it follows that, if $f(x, y, y', \ldots, y^{(n-1)})$ is continuous on $(a, b) \times R^n$ and $a < x_1 < x_2 < \cdots < x_k < b$, then $y(x) \in C^{(n)}[x_1, x_k]$ and is a solution of the $k$-point boundary value problem

$$y^{(n)} = f(x, y, y', \ldots, y^{(n-1)}), \tag{1.1}$$

$$y^{(i)}(x_j) = c_{ji},\ 0 \leqslant i \leqslant m_j - 1,\ 1 \leqslant j \leqslant k, \tag{1.2}$$

if and only if $y(x) \in C^{(n-1)}[x_1, x_k]$ and is a solution of the integral equation

$$y(x) = \int_{x_1}^{x_k} G(x,s) f(s, y(s), y'(s), \ldots, y^{(n-1)}(s))\, ds + \omega(x), \tag{2.16}$$

where $G(x,s)$ is the Green's function for the associated problem (2.12), (2.13) and $\omega(x)$ is the solution of $y^{(n)} = 0$ satisfying (1.2).

Let $k$ with $2 \leqslant k \leqslant n$ be fixed, let the integers $m_j$, $1 \leqslant j \leqslant k$, be fixed, and let $G(x,s)$ be the Green's function for the corresponding boundary value problem (2.12), (2.13). Then there are constants $\gamma_i$, $0 \leqslant i \leqslant n-1$, which depend only on $k$ and on the integers $m_j$, $1 \leqslant j \leqslant k$, and not on the spacing of the points $x_1 < x_2 < \cdots < x_k$, and are such that

$$\int_{x_1}^{x_k} \left| \frac{\partial^i}{\partial x^i} G(x,s) \right| ds \leqslant \gamma_i (x_k - x_1)^{n-i} \tag{2.17}$$

for $x_1 \leqslant x \leqslant x_k$ and $0 \leqslant i \leqslant n-1$. Suitable constants $\gamma_i$ can be found by direct calculation for small integer values of $n$ or for certain $k$-point problems for arbitrary $n$, for example, for $n$-point problems. For arbitrary $n$ and arbitrary $k$ the calculations become difficult to manage. Also it is frequently very difficult to determine the smallest values of the constants $\gamma_i$. Estimates on Green's functions for $k$-point boundary value problems for $y^{(n)} = 0$ have been given considerable attention, for example, Wend [4], Beesack [5], Nehari [6], Das and Vatsala [7], and Gustafson [8].

With regard to the boundary value problem

$$y^{(n)} = f(x, y, y', \ldots, y^{(n-1)}), \tag{1.1}$$

$$y^{(i)}(x_j) = c_{ji},\ 0 \leqslant i \leqslant m_j - 1,\ 1 \leqslant j \leqslant k, \tag{1.2}$$

we assume in the rest of this section that we are dealing with a particular type of $k$-point problem, that is, $2 \leqslant k \leqslant n$ fixed and $m_j$, $1 \leqslant j \leqslant k$, fixed. The associated Green's function for $y^{(n)} = 0$ satisfies the inequalities (2.17) with suitable values for the constants $\gamma_i$, $0 \leq i \leq n-1$. Then the following results, which are analogous to those obtained for 2-point boundary value problems for second order equations, can be proven:

THEOREM 2.4: *Assume that $f(x, y, y', \ldots, y^{(n-1)})$ is continuous on $[a, b] \times R^n$ and satisfies the Lipschitz condition*

$$|f(x, y_1, y_2, \ldots, y_n) - f(x, z_1, z_2, \ldots, z_n)| \leq \sum_{i=1}^{n} P_{i-1}|y_i - z_i|$$

*on $[a, b] \times R^n$. Then for arbitrary values of the boundary values $c_{ji}$, $0 \leq i \leq m_j - 1$, $1 \leq j \leq k$, the boundary value problem* (1.1), (1.2) *has a unique solution provided $a \leq x_1 < x_2 < \cdots < x_k \leq b$ and $x_k - x_1 < \delta$, where $\delta$ is the positive root of the equation*

$$\sum_{i=1}^{n} \gamma_{i-1} P_{i-1} \delta^{n-i+1} = 1.$$

THEOREM 2.5: *Assume that $f(x, y, y', \ldots, y^{(n-1)})$ is continuous on $[a, b] \times R^n$ and let $N_i$, $0 \leq i \leq n-1$, be given positive constants. Then there is a $\delta = \delta(N_0, N_1, \ldots, N_{n-1}) > 0$ such that the boundary value problem* (1.1), (1.2) *has a solution provided $a \leq x_1 < x_2 < \cdots < x_k \leq b$, $x_k - x_1 \leq \delta$, and $|\omega^{(i)}(x)| \leq N_i$ on $[x_1, x_k]$ for $0 \leq i \leq n-1$ where $\omega(x)$ is the solution of $y^{(n)} = 0$ which satisfies the boundary conditions* (1.2). *In fact we may take*

$$\delta(N_0, N_1, \ldots, N_{n-1}) = \operatorname{Min}\left\{\left(\frac{N_i}{\gamma_i Q}\right)^{\frac{1}{n-i}} \middle|\, 0 \leq i \leq n-1\right\},$$

*where*

$$Q = \operatorname{Max}\{|f(x, y, y', \ldots, y^{(n-1)})|\,|a \leq x \leq b, |y^{(i)}| \leq 2N_i$$
$$\text{for } 0 \leq i \leq n-1\}.$$

COROLLARY 2.6: *If $f(x, y, y', \ldots, y^{(n-1)})$ is bounded and continuous on $[a, b] \times R^n$, then for any $a \leq x_1 < x_2 < \cdots < x_k \leq$*

*b and any choice of the boundary values* $c_{ji}$, $1 \leqslant j \leqslant k$, $0 \leqslant i \leqslant m_j - 1$, *the boundary value problem* (1.1), (1.2) *has a solution.*

The proofs of Theorem 2.4, Theorem 2.5, and Corollary 2.6, which differ only slightly from the proofs of Theorem 2.1, Theorem 2.2, and Corollary 2.3, will be omitted.

The following additional corollary of Theorem 2.5 is frequently useful:

COROLLARY 2.7: *Assume that* $f(x, y, y', \ldots, y^{(n-1)})$ *is continuous on* $[a, b] \times R^n$ *and that* $g(x)$ *is a given function with* $g \in C^{(n-1)}[a, b]$. *Then there is a* $\delta > 0$, *depending on* $g(x)$, *such that for* $a \leqslant x_1 < x_2 < \cdots < x_k \leqslant b$ *and* $x_k - x_1 < \delta$ *the boundary value* (1.1), (1.2) *has a solution where in* (1.2) *the boundary conditions are given by* $y^{(i)}(x_j) = g^{(i)}(x_j)$ *for* $1 \leqslant j \leqslant k$, $0 \leqslant i \leqslant m_j - 1$.

*Proof:* In this case $\omega(x)$, the polynomial of degree $n-1$, satisfies $\omega^{(i)}(x_j) = g^{(i)}(x_j)$ for $1 \leqslant j \leqslant k$, $0 \leqslant i \leqslant m_j - 1$, and it follows from repeated applications of Rolle's Theorem that there is an $x_0$, $x_1 < x_0 < x_k$, such that $\omega^{(n-1)}(x_0) = g^{(n-1)}(x_0)$. Since $\omega(x)$ is a polynomial of degree $n-1$, it follows that $\omega^{(n-1)}(x) \equiv g^{(n-1)}(x_0)$ on $[x_1, x_k]$. Furthermore, for each $0 \leqslant i \leqslant n-2$ there is at least one point in $[x_1, x_k]$ at which $\omega^{(i)}(x) = g^{(i)}(x)$. Thus, if $\tau \in [x_1, x_k]$ is such that $\omega^{(n-2)}(\tau) = g^{(n-2)}(\tau)$, we have

$$\begin{aligned}\omega^{(n-2)}(x) &= \omega^{(n-2)}(\tau) + \int_\tau^x \omega^{(n-1)}(s)\, ds \\ &= g^{(n-2)}(\tau) + \int_\tau^x g^{(n-1)}(x_0)\, ds\end{aligned}$$

on $[x_1, x_k]$. It follows that, if $|g^{(i)}(x)| \leqslant M_i$ on $[a, b]$ for $0 \leqslant i \leqslant n-1$, then $|\omega^{(n-1)}(x)| \leqslant M_{n-1}$ and $|\omega^{(n-2)}(x)| \leqslant M_{n-2} + M_{n-1}(b-a)$ on $[x_1, x_k]$. Using the same argument repeatedly, we obtain $|\omega^{(i)}(x)| \leqslant N_i$ on $[x_1, x_k]$ for $0 \leqslant i < n-1$ where

$$N_i = \sum_{j=i}^{n-1} M_j (b-a)^{j-i}.$$

Thus, with these values of $N_i$, $0 \leqslant i \leqslant n-1$, it follows from Theorem 2.5 that the boundary value problem has a solution provided $a \leqslant x_1 < x_2 < \cdots < x_k \leqslant b$ and $x_k - x_1 < \delta = \delta(N_0, N_1, \ldots, N_{n-1})$.

## 3. APPLICATIONS OF DIFFERENTIAL INEQUALITIES

In this Section we consider boundary value problems from the global standpoint and consider first the case in which all solutions of the differential equation may not extend to the interval of concern in the given boundary value problem. It is then necessary to show that there is at least one solution that does extend and does satisfy the specified boundary conditions. In the case of second order equations this result is achieved by assuming the existence of solutions of differential inequalities that behave in a suitable way and by assuming that solutions of the differential equation either extend or become unbounded. Corresponding theorems for higher order equations impose additional monotoneity conditions on the function $f(x, y, y', \ldots, y^{(n-1)})$ in the differential equation. The arguments employed require a form of the basic Kamke convergence theorem for solutions of initial value problems for ordinary differential equations. A more general form of this most important theorem may be found in most advanced treatises on differential equations, for example, [1, p. 14].

KAMKE THEOREM: *Assume that in the equations*

$$y^{(n)} = f_k(x, y, y', \ldots, y^{(n-1)}), \quad k = 0, 1, 2, \ldots, \tag{3.1}$$

*the functions* $f_k(x, y, y', \ldots, y^{(n-1)})$ *are continuous on* $I \times R^n$ *where* $I$ *is an interval of the reals and assume that* $\lim_{k\to\infty} f_k(x, y, y', \ldots, y^{(n-1)}) = f_0(x, y, y', \ldots, y^{(n-1)})$ *uniformly on each compact subset of* $I \times R^n$. *Assume that* $\{x_k\}_{k=0}^{\infty} \subset I$ *with* $\lim_{k\to\infty} x_k = x_0$ *and that, for each integer* $k \geqslant 1$, $y_k(x)$ *is a solution of* $(3.1)_k$ *which is defined on a maximal interval* $I_k \subset I$ *with* $x_k \in I_k$ *for each* $k \geqslant 1$. *Further, assume that* $\lim_{k\to\infty} y_k^{(i-1)}(x_k) = y_i$ *for each* $1 \leqslant i \leqslant n$. *Then there is a subsequence* $\{y_{k_j}(x)\}$ *of* $\{y_k(x)\}$ *and there is a solution* $y_0(x)$ *of* $(3.1)_0$ *defined on a maximal interval* $I_0 \subset I$ *such that* $x_0 \in I_0$, $y_0^{(i-1)}(x_0) = y_i$ *for each* $1 \leqslant i \leqslant n$, *and such that for any compact interval* $[c, d] \subset I_0$ *it follows that* $[c, d] \subset I_{k_j}$ *for all sufficiently large* $k_j$ *and* $\lim y_{k_j}^{(i-1)}(x) = y_0^{(i-1)}(x)$ *uniformly on* $[c, d]$ *for each* $1 \leqslant i \leqslant n$.

DEFINITION 3.1: A function $\psi(x)$ is said to be an upper solution of $y^{(n)} = f(x, y, \ldots, y^{(n-1)})$ on an interval $[a, b]$ in case $\psi \in C^{(n)}[a, b]$ and

$$\psi^{(n)}(x) \leqslant f\big(x, \psi(x), \psi'(x), \ldots, \psi^{(n-1)}(x)\big) \tag{3.2}$$

on $[a, b]$. Similarly, $\phi(x)$ is said to be a lower solution in case $\phi \in C^{(n)}[a, b]$ and

$$\phi^{(n)}(x) \geqslant f\big(x, \phi(x), \phi'(x), \ldots, \phi^{(n-1)}(x)\big) \tag{3.3}$$

on $[a, b]$.

We consider first the 2-point boundary value problem for a second order equation

$$y'' = f(x, y, y'), \tag{3.4}$$

$$y(a) = A, y(b) = B. \tag{3.5}$$

THEOREM 3.2: *Assume that* $f(x, y, y')$ *is continuous on* $[a, b] \times R^2$ *and that each solution of* (3.4) *either extends to* $[a, b]$ *or becomes unbounded on its maximal interval of existence. Assume that there exist an upper solution* $\psi(x)$ *and a lower solution* $\phi(x)$ *of* (3.4) *on* $[a, b]$ *such that* $\phi(x) \leqslant \psi(x)$ *on* $[a, b]$, $\phi(a) \leqslant A \leqslant \psi(a)$, *and* $\phi(b) \leqslant B \leqslant \psi(b)$. *Then the boundary value problem* (3.4), (3.5) *has a solution* $y(x)$ *with* $\phi(x) \leqslant y(x) \leqslant \psi(x)$ *on* $[a, b]$.

*Proof:* First define $F(x, y, y')$ on $[a, b] \times R^2$ by

$$F(x,y,y') = \begin{cases} f(x, \psi(x), y') + \dfrac{y - \psi(x)}{1 + y - \psi(x)}, & \text{for } y > \psi(x), \\ f(x, y, y'), & \text{for } \phi(x) \leqslant y \leqslant \psi(x), \\ f(x, \phi(x), y') + \dfrac{y - \phi(x)}{1 + |y - \phi(x)|}, & \text{for } y < \phi(x). \end{cases}$$

Let $N > 0$ be such that $|\phi'(x)| \leqslant N$ and $|\psi'(x)| \leqslant N$ on $[a, b]$. Then define the functions $F_k(x, y, y')$ for $k = 1, 2, 3, \ldots,$ by

$$F_k(x,y,y') = \begin{cases} F(x, y, N + k), & \text{for } y' > N + k, \\ F(x, y, y'), & \text{for } |y'| \leqslant N + k, \\ F(x, y, -N - k) & \text{for } y' < -N - k. \end{cases}$$

Then for each $k \geqslant 1$ $F_k(x, y, y')$ is continuous and bounded on $[a, b] \times R^2$ and $\lim_{k\to\infty} F_k(x, y, y') = F(x, y, y')$ uniformly on each compact subset of $[a, b] \times R^2$.

Since each $F_k(x, y, y')$ is bounded and continuous, it follows from Corollary 2.3 that the boundary value problem

$$y'' = F_k(x, y, y'),$$

$$y(a) = A, y(b) = B,$$

has a solution $y_k(x)$ for each integer $k \geqslant 1$. We claim that $\phi(x) \leqslant y_k(x) \leqslant \psi(x)$ on $[a, b]$ for each $k \geqslant 1$. To see this assume that for some $y_k(x)$ it is true that $y_k(x) > \psi(x)$ at some points in $[a, b]$. Then, since $y_k(x) \leqslant \psi(x)$ at $x = a$ and $x = b$, it follows that $y_k(x) - \psi(x)$ has a positive maximum at some $x_0$ with $a < x_0 <$

$b$. This implies that

$$y_k''(x_0) - \psi''(x_0) \leqslant 0 \quad \text{and} \quad y_k'(x_0) = \psi'(x_0).$$

Thus

$$|y_k'(x_0)| = |\psi'(x_0)| \leqslant N < N + k$$

and

$$F_k(x_0, y_k(x_0), y_k'(x_0)) = F(x_0, y_k(x_0), y_k'(x_0))$$

by the definition of $F_k$. Furthermore, since $y_k(x_0) > \psi(x_0)$ and $y_k'(x_0) = \psi'(x_0)$, we have

$$F(x_0, y_k(x_0), y_k'(x_0)) = f(x_0, \psi(x_0), \psi'(x_0)) + \frac{y_k(x_0) - \psi(x_0)}{1 + y_k(x_0) - \psi(x_0)}.$$

Coupling this with the fact that $\psi(x)$ is an upper solution, we have

$$y_k''(x_0) - \psi''(x_0) \geqslant \frac{y_k(x_0) - \psi(x_0)}{1 + y_k(x_0) - \psi(x_0)} > 0.$$

From this contradiction we conclude that $y_k(x) \leqslant \psi(x)$ on $[a, b]$. In a similar way it can be shown that $y_k(x) \geqslant \phi(x)$ on $[a, b]$.

Since $\phi(x) \leqslant y_k(x) \leqslant \psi(x)$ on $[a, b]$ for each $k$, it follows that for each $k$ there is an $x_k \in (a, b)$ such that

$$(b - a)|y_k'(x_k)| = |y_k(b) - y_k(a)| \leqslant \operatorname{Max}\{|\psi(a) - \phi(b)|, |\psi(b) - \phi(a)|\}.$$

Thus, each of the sequences $\{y_k(x_k)\}$ and $\{y_k'(x_k)\}$ is bounded and it follows that we can choose a subsequence, which we shall renumber as the original sequence, such that $\lim x_k = x_0$, $\lim y_k(x_k) = y_0$, and $\lim y_k'(x_k) = y_1$. It follows from the Kamke Theorem that there is a solution $y(x)$ of $y'' = F(x, y, y')$ on a maximal subinterval $I \subset [a, b]$ and there is a subsequence of $\{y_k(x)\}$ which converges uniformly to $y(x)$ on each compact subinterval of $I$. Hence, $\phi(x) \leqslant y(x) \leqslant \psi(x)$ for all $x \in I$. Since $y(x)$ extends to $[a, b]$ or becomes unbounded on $I$ if $I$ is a proper subinterval of $[a, b]$, we conclude that $I = [a, b]$ and that a subsequence of $\{y_k(x)\}$ converges to $y(x)$ uniformly on $[a, b]$. It follows that $y(a) = A$ and $y(b) = B$. Since $\phi(x) \leqslant y(x) \leqslant \psi(x)$ on $[a, b]$, it follows from the definition of $F(x, y, y')$ that $y(x)$ is a solution of (3.4) on $[a, b]$.

The above proof of Theorem 3.2 is essentially contained in [9] and in [10].

It is important to know conditions in terms of the function $f(x, y, y')$ in (3.4) which are sufficient to guarantee that each solution either extends to $[a, b]$ or becomes unbounded on its maximal interval of existence. The most well known such condition is due to Nagumo [11].

DEFINITION 3.3: The equation (3.4) is said to satisfy a Nagumo condition on $[a, b] \times R^2$ in case $f(x, y, y')$ is continuous on $[a, b] \times R^2$ and given any $M > 0$ there is a positive continuous function $h_M(s)$ on $[0, \infty)$ such that

$$|f(x, y, y')| \leqslant h_M(|y'|)$$

for all $(x, y, y') \in [a, b] \times R^2$ with $|y| \leqslant M$ and such that

$$\int_0^\infty \frac{s\, ds}{h_M(s)} = +\infty.$$

THEOREM 3.4: *If equation* (3.4) *satisfies a Nagumo condition on* $[a, b] \times R^2$, *then* (3.4) *has the property that each solution either extends to* $[a, b]$ *or becomes unbounded on its maximal interval of existence.*

The proof of Theorem 3.4 may be found in [1, p. 428] or in [12, p. 353].

Versions of Theorem 3.2 with much more general boundary conditions have been proven. For example, in [13] boundary conditions of the form $g(y(a), y'(a)) = 0$, $h(y(b), y'(b)) = 0$ are considered where the functions $g(y, y')$ and $h(y, y')$ are continuous on $[\phi(a), \psi(a)] \times R$ and $[\phi(b), \psi(b)] \times R$, respectively, and satisfy certain monotoneity conditions in $y'$. Also, versions of Theorem 3.2 with more general boundary conditions have been proven using Lyapunov functions or variations of the topological method of Wazewski, for example [14], [15], [16] and [17].

Theorems similar to Theorem 3.2 have been proven for equations of higher order, for example, Klaasen [10] has proven the following theorem for the boundary value problem

$$y''' = f(x, y, y', y''), \tag{3.6}$$

$$y(a) = \alpha, y'(a) = \beta, y'(b) = \gamma. \tag{3.7}$$

THEOREM 3.5: *Assume that* $f(x, y, y', y'')$ *is continuous on* $[a, b] \times R^3$, *that* $f(x, y, y', y'')$ *is nonincreasing in* $y$ *for each fixed* $x, y', y''$, *and that each solution of* (3.6) *either extends to* $[a, b]$ *or becomes unbounded on its maximal interval of existence. Assume that* $\phi(x)$ *and* $\psi(x)$ *are respectively lower and upper solutions of* (3.6) *on* $[a, b]$ *such that* $\phi(x) \leqslant \psi(x)$ *and* $\phi'(x) \leqslant \psi'(x)$ *on* $[a, b]$ *and such that* $\phi(a) \leqslant \alpha \leqslant \psi(a)$, $\phi'(a) \leqslant \beta \leqslant \psi'(a)$, *and* $\phi'(b) \leqslant \gamma \leqslant \psi'(b)$. *Then the boundary value problem* (3.6), (3.7) *has a solution* $y(x)$ *with* $\phi(x) \leqslant y(x) \leqslant \psi(x)$ *and* $\phi'(x) \leqslant y'(x) \leqslant \psi'(x)$.

The proof of Theorem 3.5, although more difficult, resembles that of Theorem 3.2 and will be omitted.

Kelley [18] has shown that various extensions of Theorem 3.5 are valid for equations (1.1) of arbitrary order $n$. These results, in most cases, include the hypothesis that solutions of (1.1) either extend or become unbounded on their maximal intervals of existence. Consequently, this type of behavior of solutions of (1.1) seems to be of sufficient utility to make it worth while to investigate conditions on $f(x, y, y', \ldots, y^{(n-1)})$ sufficient to guarantee it. The following theorem, found in [19], is a result in this direction:

THEOREM 3.6: *Assume that* $f(x, y, y', \ldots, y^{(n-1)})$ *is continuous on* $I \times R^n$, *where* $I$ *is an interval of the reals, and that correspond-*

*ing to each compact interval $[c, d] \subset I$ and each positive number $M$ there exist numbers $h, p_i$, $1 \leqslant i \leqslant n-1$, such that $h > 0, p_i \geqslant 0$ for $1 \leqslant i \leqslant n-1$, $\sum_{i=1}^{n-1} ip_i < n$, and such that*

$$|f(x, y, y', \ldots, y^{(n-1)})| \leqslant h \prod_{i=1}^{n-1} \phi_{p_i}(y^{(i)})$$

*for all $(x, y, y', \ldots, y^{(n-1)})$ satisfying $c \leqslant x \leqslant d$, $|y| \leqslant M$, where $\phi_p(s) = \mathrm{Max}\{1, |s|^p\}$. Then each solution of $y^{(n)} = f(x, y, y', \ldots, y^{(n-1)})$ either extends to $I$ or becomes unbounded on its maximal interval of existence.*

## 4. RELATIONSHIPS BETWEEN UNIQUENESS AND EXISTENCE OF SOLUTIONS

In this section we are again concerned with the question of the global existence of solutions of $k$-point boundary value problems. However, we will now assume that we are not troubled by questions of extendability of solutions; that is, we will assume that all solutions of the differential equation extend throughout the interval. With this basic assumption we will then consider additional conditions under which the uniqueness of solutions of boundary value problems will imply the existence of solutions of those problems. Most of our discussion will be confined to the $n$-point boundary value problem

$$y^{(n)} = f(x, y, y', \ldots, y^{(n-1)}), \tag{4.1}$$

$$y(x_j) = c_j,\ 1 \leqslant j \leqslant n. \tag{4.2}$$

As observed in Section 1, if equation (4.1) is a linear equation of the form

$$y^{(n)} = \sum_{i=0}^{n-1} p_i(x) y^{(i)} + q(x) \tag{4.3}$$

with the coefficient functions $q(x)$ and $p_i(x)$, $0 \leqslant i \leqslant n-1$, continuous on an interval $(a, b)$, and if $x_1 < x_2 < \cdots < x_n$ is some fixed set of points in the interval $(a, b)$, then the uniqueness of solutions of boundary value problems (4.3), (4.2) implies the existence of solutions for all choices of the boundary values $c_j$, $1 \leqslant j \leqslant n$. Again as observed in Section 1, the argument employed in proving this assertion is algebraic and is based on the linear structure of the solution set of (4.3) and on the linearity of the boundary conditions (4.2). Lasota and Opial [20] in 1967 published the first result establishing that uniqueness implies existence of solutions of boundary value problems for nonlinear differential equations. Their result is contained in Theorem 4.2 below.

In the proof of Theorem 4.2 and in our subsequent discussion we will need the following theorem concerning the continuity of solutions of initial value problems with respect to initial conditions:

THEOREM 4.1: *Assume that for the equation*

$$y^{(n)} = f(x, y, y', \ldots, y^{(n-1)}) \tag{4.1}$$

*the following conditions are satisfied*:

(A) $f(x, y, y', \ldots, y^{(n-1)})$ *is continuous on* $I \times R^n$ *where* $I$ *is an interval*,

(B) *All solutions of* (4.1) *extend to* $I$, *and*

(C) *Solutions of initial value problems for* (4.1) *are unique.*

*Then solutions of* (4.1) *depend continuously on initial conditions in the sense that, given any solution* $y(x)$ *of* (4.1), *given any* $x_0 \in I$, *given any compact interval* $[c, d] \subset I$, *and given any* $\varepsilon > 0$, *there is a* $\delta > 0$ *such that, if* $z(x)$ *is any solution of* (4.1) *with* $|z^{(i)}(x_0) - y^{(i)}(x_0)| < \delta$ *for* $0 \leqslant i \leqslant n-1$, *then* $|y^{(i)}(x) - z^{(i)}(x)| < \varepsilon$ *on* $[c, d]$ *for* $0 \leqslant i \leqslant n-1$.

The proof of this Theorem follows directly from the Kamke Convergence Theorem of Section 3.

THEOREM 4.2: *Assume that the equation*

$$y'' = f(x, y, y') \tag{4.4}$$

*satisfies the conditions* (A), (B), *and* (C) *of Theorem* 4.1 *on the slab* $[a, b) \times R^2$. *In addition assume that the following uniqueness condition for 2-point boundary value problems is satisfied:*

(D) *For any* $a \leqslant x_1 < x_2 < b$ *and any solutions* $y(x)$ *and* $z(x)$ *of* (4.4), $y(x_i) = z(x_i)$ *for* $i = 1, 2$ *implies* $y(x) \equiv z(x)$. *Then for any* $a \leqslant x_1 < x_2 < b$ *and any* $y_1, y_2 \in R$ *the 2-point boundary value problem*

$$y'' = f(x, y, y'), \tag{4.4}$$

$$y(x_1) = y_1, y(x_2) = y_2, \tag{4.5}$$

*has a solution.*

*Proof:* Let $a \leqslant x_1 < x_2 < b$ and $y_1, y_2 \in R$ be given and let $y(x; m)$ be the solution of (4.4) satisfying the initial conditions $y(x_1) = y_1, y'(x_1) = m$. Then to show that (4.4), (4.5) has a solution it suffices to show that $S \equiv \{y(x_2; m) | -\infty < m < +\infty\}$ is the whole real line. It follows from Theorem 4.1 that $y(x_2; m)$ is a continuous function of $m$ which in turn implies that $S$ is an interval. Consequently, to show that $S = R$ it suffices to show that $S$ is neither bounded above nor below. Assume that $S$ is bounded above and that $\beta \in R$ is such that $y(x_2; m) < \beta$ for all $m \in R$. Let $u(x)$ be the solution of (4.4) satisfying the initial conditions $u(x_2) = \beta$, $u'(x_2) = 0$. Since $\beta \notin S$, $u(x_1) \neq y_1$. Assume first that $u(x_1) > y_1$ and let $\{y_n(x)\}_{n=1}^{\infty}$ be the sequence of solutions of (4.4) defined by $y_n(x) = y(x; n)$. Then it follows from the uniqueness hypothesis (D) that

$$y_1(x) \leqslant y_n(x) \leqslant u(x)$$

on $[x_1, x_2]$ for all $n \geqslant 1$. Thus, if $M > 0$ is such that $|y_1(x)| \leqslant M$ and $|u(x)| \leqslant M$ on $[x_1, x_2]$, then $|y_n(x)| \leqslant M$ on $[x_1, x_2]$ and for each $n \geqslant 1$ there is a $t_n \in (x_1, x_2)$ such that $(x_2 - x_1)|y_n'(t_n)| = |y_n(x_2) - y_n(x_1)| \leqslant 2M$. Consequently, we may select a subsequence, relabeled as the original sequence, such that $\{t_n\}$, $\{y_n(t_n)\}$, and $\{y_n'(t_n)\}$ all converge. Then, by the Kamke Theorem and hypothesis (B) there is a further subsequence such that $\{y_n^{(i)}(x)\}$ converges uniformly on $[x_1, x_2]$ for $i = 0, 1$. However, this is impossible since $y_n'(x_1) = y'(x_1; n) = n$. We conclude that it is not possible to have $u(x_1) > y_1$ and therefore $u(x_1) < y_1$. In this case, if $\{y_n(x)\}_{n=1}^{\infty}$ is again the sequence with $y_n(x) = y(x; n)$ for each $n$ and if $x_2 < x_3 < b$, then $y_1(x) \leqslant y_n(x) \leqslant u(x)$ on $[x_2, x_3]$. This leads to the same contradiction as before which now forces us to the conclusion that $S$ cannot be bounded above. A similar argument leads to the conclusion that $S$ is not bounded below. Thus, $S = R$ and the boundary value problem (4.4), (4.5) has a solution.

With more involved arguments Theorem 4.1 can be proven without the assumption that solutions of initial value problems are unique. The proof of Theorem 4.2 made use of the availability of an interval on each side of $x = x_2$ and in fact the Theorem is false when stated in terms of a closed interval $[a, b]$, [12, p. 347].

A number of generalizations of Theorem 4.2 have been proven both in terms of generalizing the boundary conditions and in terms of weakening the uniqueness hypothesis for solutions of boundary value problems, for example, [21], [22], [23] and [24].

We turn now to the question of the possibility of extending Theorem 4.2 to $n$-point boundary value problems for the equation

$$y^{(n)} = f(x, y, y', \ldots, y^{(n-1)}). \tag{4.1}$$

This was done in [25] for equations of order 3 and was done independently for equations of arbitrary order by Hartman [26] and Klaasen [27]. The arguments we shall use are slight modifications of those used by Hartman.

To simplify our arguments we will assume that (4.1) satisfies the Hypotheses (A), (B) and (C) of Theorem 4.1 on a slab $(a, b) \times R^n$. However, the results obtained are valid if the hypotheses are stated in terms of slabs on half-open intervals. Our proof of the theorem which extends Theorem 4.2 to equations of arbitrary order requires the existence of solutions of $n$-point boundary value problems near a given solution for equation (4.1) and the continuity of solutions with respect to boundary conditions. These results in turn will be proven by using the Brouwer theorem on the invariance of domain [28, p. 199].

BROUWER THEOREM: *If $U$ is an open subset of $R^n$, $n$-dimensional Euclidean space, and $\phi : U \to R^n$ is one-to-one and continu-*

*ous on $U$, then $\phi$ is a homeomorphism and $\phi(U)$ is an open subset of $R^n$.*

THEOREM 4.3: *Assume that the equation*

$$y^{(n)} = f(x, y, y', \ldots, y^{(n-1)}) \tag{4.1}$$

*satisfies the following conditions:*

(A) *$f(x, y, y', \ldots, y^{(n-1)})$ is continuous on $(a, b) \times R^n$.*

(B) *All solutions of* (4.1) *extend to $(a, b)$.*

(C) *Solutions of initial value problems for* (4.1) *are unique.*

*and*

(D) *For any $a < x_1 < x_2 < \cdots < x_n < b$ and any solutions $y(x)$ and $z(x)$ of* (4.1), *$y(x_i) = z(x_i)$ for $1 \leqslant i \leqslant n$ implies $y(x) \equiv z(x)$ on $(a, b)$.*

*Then, given any $a < x_1 < x_2 < \cdots < x_n < b$ and any solution $y(x)$ of* (4.1), *there is an $\varepsilon > 0$ such that $|t_i - x_i| < \varepsilon$ and $|y(x_i) - y_i| < \varepsilon$ for $1 \leqslant i \leqslant n$ implies that there is a solution $z(x)$ of* (4.1) *with $z(t_i) = y_i$ for $1 \leqslant i \leqslant n$. Furthermore, if for each $i$, $1 \leqslant i \leqslant n$, $\{t_{ik}\}_{k=1}^{\infty} \subset (a, b)$ is a sequence with $\lim_{k\to\infty} t_{ik} = x_i$ and if $\{z_k(x)\}_{k=1}^{\infty}$ is a sequence of solutions of* (4.1) *such that $\lim_{k\to\infty} z_k(t_{ik}) = y(x_i)$ for each $1 \leqslant i \leqslant n$, then $\lim_{k\to\infty} z_k^{(j)}(x) = y^{(j)}(x)$ uniformly on compact subintervals of $(a, b)$ for each $0 \leqslant j \leqslant n - 1$.*

*Proof:* Let $\Delta = \{(x_1, x_2, \ldots, x_n) | a < x_1 < x_2 < \cdots < x_n < b\}$. Then $\Delta \times R^n$ is an open subset of $R^{2n}$. Let $x_0$ be an arbitrary but fixed point in $(a, b)$ and define $\phi : \Delta \times R^n \to R^{2n}$ by

$$\phi(x_1, x_2, \ldots, x_n, c_1, c_2, \ldots, c_n)$$

$$= (x_1, x_2, \ldots, x_n, y(x_1), y(x_2), \ldots, y(x_n)), \tag{4.6}$$

where $y(x)$ is the unique solution of (4.1) satisfying the initial conditions $y^{(i-1)}(x_0) = c_i$, $1 \leqslant i \leqslant n$. Then it follows from Theorem 4.1 that $\phi$ is continuous on $\Delta \times R^n$. Assume that $(x_1, \ldots, x_n, c_1, \ldots, c_n)$ and $(t_1, \ldots, t_n, d_1, \ldots, d_n)$ in $\Delta \times R^n$

are such that

$$\phi(x_1, \ldots, x_n, c_1, \ldots, c_n) = \phi(t_1, \ldots, t_n, d_1, \ldots, d_n).$$

It follows from the definition of $\phi$ in (4.6) that $x_i = t_i$ for $1 \leqslant i \leqslant n$ and that $y(x_i) = z(x_i)$ for $1 \leqslant i \leqslant n$ where $y(x)$ and $z(x)$ are the solutions of (4.1) with $y^{(i-1)}(x_0) = c_i$ and $z^{(i-1)}(x_0) = d_i$ for $1 \leqslant i \leqslant n$. However, by Hypothesis (D) the condition $y(x_i) = z(x_i)$ for $1 \leqslant i \leqslant n$ implies $y(x) \equiv z(x)$ on $(a, b)$. Thus we also have $c_i = d_i$ for $1 \leqslant i \leqslant n$ and $\phi$ is one-to-one on $\Delta \times R^n$. It follows from the Brouwer Theorem that $\phi(\Delta \times R^n)$ is an open set in $R^{2n}$ and that $\phi^{-1}$ is continuous on $\phi(\Delta \times R^n)$. The first assertion of the Theorem follows from the fact that $\phi(\Delta \times R^n)$ is open. The second assertion, that is the continuity of solutions with respect to boundary values, follows from the continuity of $\phi^{-1}$ and the continuity of solutions of (4.1) with respect to initial values as asserted in Theorem 4.1.

COROLLARY 4.4: *Assume that equation* (4.1) *satisfies the Hypotheses* (A), (B), (C) *and* (D) *of Theorem* 4.3. *Then, if* $a < x_1 < x_2 < \cdots < x_{n-1} < b$ *and if* $y(x)$ *and* $z(x)$ *are distinct solutions of* (4.1) *such that* $y(x_i) = z(x_i)$ *for* $1 \leqslant i \leqslant n-1$, *it follows that* $y(x) - z(x)$ *changes sign at each* $x_i$, $1 \leqslant i \leqslant n$.

*Proof:* Since $y(x)$ and $z(x)$ are assumed to be distinct solutions, $y(x) - z(x)$ is zero only at the points $x_i$, $1 \leqslant i \leqslant n-1$. Assume that for some $j$, $1 \leqslant j \leqslant n-1$, $y(x) - z(x)$ has the same sign on each side of $x_j$, say $y(x) - z(x) > 0$ on right and left open intervals adjacent to $x_j$. Choose an $x_0 \in (a, b)$ with $x_0 \neq x_i$ for $1 \leqslant i \leqslant n-1$. Then it follows from Theorem 4.3 that for $\varepsilon > 0$ sufficiently small (4.1) has a solution $y(x; \varepsilon)$ with $y(x_i; \varepsilon) = y(x_i)$ for $0 \leqslant i \leqslant n-1$, $i \neq j$, and $y(x_j; \varepsilon) = y(x_j) - \varepsilon$. Furthermore, $\lim_{\varepsilon \to 0} y(x; \varepsilon) = y(x)$ uniformly on compact subintervals of $(a, b)$. Thus, if $\delta > 0$ is chosen so that $[x_j - \delta, x_j + \delta] \subset (a, b)$ and $x_i \notin [x_j - \delta, x_j + \delta]$ for $i \neq j$, then for sufficiently small $\varepsilon > 0$ we will have $y(x_j \pm \delta; \varepsilon) > z(x_j \pm \delta)$ and $y(x_j, \varepsilon) = z(x_j) - \varepsilon$. Hence, $y(x; \varepsilon) - z(x)$ will have distinct zeros at $x = x_i$, $1 \leqslant i \leqslant n-1$, $i \neq j$, and two distinct zeros in $(x_j - \delta, x_j + \delta)$. This contradicts (D) and we conclude that $y(x) - z(x)$ must change sign at $x_j$.

We are now ready to prove the theorem which is the extension of Theorem 4.2 to equations of arbitrary order.

THEOREM 4.5: *Assume that the equation*

$$y^{(n)} = f(x, y, y', \ldots, y^{(n-1)}) \tag{4.1}$$

*satisfies the Hypotheses* (A), (B), (C) *and* (D) *of Theorem* 4.3 *and the additional Hypothesis*:

(E) *If* $[c, d]$ *is a compact interval of* $(a, b)$ *and* $\{y_k(x)\}_{k=1}^{\infty}$ *is a sequence of solutions of* (4.1) *such that* $|y_k(x)| \leqslant M$ *on* $[c, d]$ *for some* $M > 0$ *and all* $k = 1, 2, 3, \ldots,$ *then there is a subsequence* $\{y_{k_j}(x)\}$ *such that* $\{y_{k_j}^{(i)}(x)\}$ *converges uniformly on* $[c, d]$ *for each* $0 \leqslant i \leqslant n-1$.

*Then for any* $a < x_1 < x_2 < \cdots < x_n < b$ *and any real numbers* $y_i$, $1 \leqslant i \leqslant n$, *the n-point boundary value problem*

$$y^{(n)} = f(x, y, y', \ldots, y^{(n-1)}), \tag{4.1}$$

$$y(x_i) = y_i, \; 1 \leqslant i \leqslant n, \tag{4.2}$$

*has a solution.*

*Proof:* Let a particular boundary value problem (4.1), (4.2) be specified. Let $y_0(x)$ be an arbitrary but fixed solution of (4.1) on $(a, b)$. It will suffice to show that there is a solution $y_1(x)$ such that $y_1(x_1) = y_1$ and $y_1(x_j) = y_0(x_j)$ for $2 \leqslant j \leqslant n$. For, if this has been done, we can repeat the argument starting with $y_1(x)$ to obtain a solution $y_2(x)$ such that $y_2(x_2) = y_2$ and $y_2(x_j) = y_1(x_j)$ for $1 \leqslant j \leqslant n, j \neq 2$, and, proceeding in this way, obtain a solution of (4.1), (4.2) in $n$ steps.

Let

$$S = \{y(x_1) | y(x) \quad \text{is a solution of (4.1) and} \\ y(x_j) = y_0(x_j), \quad 2 \leqslant j \leqslant n\}. \tag{4.7}$$

Then it follows from Theorem 4.3 that $S$ is an open subset of the reals. If it can be shown that $S$ is also closed, it will follow that $S = R$ and that there is a solution $y_1(x)$ of (4.1) with $y_1(x_1) = y_1$ and $y_1(x_j) = y_0(x_j)$ for $2 \leqslant j \leqslant n$. Hence, assume $r_0 \in R$ is a limit point of $S$ which is not contained in $S$ and that $\{r_k\}_{k=1}^{\infty} \subset S$ is such that $\lim r_k = r_0$. We can assume that $\{r_k\}_{k=1}^{\infty}$ is strictly monotone and to deal with a specific case assume the sequence is strictly increasing. Let $\{y_k(x)\}$ be the sequence of solutions of (4.1) such that $y_k(x_1) = r_k$ and $y_k(x_j) = y_0(x_j)$ for $2 \leqslant j \leqslant n$. It follows from Corollary 4.4 that $\{y_k(x)\}$ is strictly increasing on $(a, x_2)$, is strictly decreasing on $(x_2, x_3)$, and is alternately strictly increasing and decreasing on the successive intervals $(x_3, x_4)$, $\ldots$ ,$(x_n, b)$.

If the sequence $\{y_k(x)\}$ is bounded on any compact subinterval of $(a, b)$ it follows from Hypothesis (E) and the Kamke Theorem that there is a solution $z(x)$ of (4.1) and a subsequence of $\{y_k(x)\}$ which converges uniformly to $z(x)$ on each compact subinterval of $(a, b)$. However, this would imply that $z(x_1) = r_0$ and $z(x_j) = y_0(x_j)$ for $2 \leqslant j \leqslant n$ which contradicts the assumption that $r_0 \notin S$. Thus $\{y_k(x)\}$ is not bounded on any compact subinterval of $(a, b)$. Let $u(x)$ be a solution of (4.1) with $u(x_1) = r_0$, for example, let $u(x)$ be the solution of (4.1) satisfying the initial conditions $u(x_1) = r_0$, $u^{(i)}(x_1) = 0$, $1 \leqslant i \leqslant n - 1$. Using the monotoneity properties of $\{y_k(x)\}$ observed above and the fact that the sequence is unbounded on each of the intervals $(a, x_1)$, $(x_1, x_2), \ldots, (x_{n-1}, x_n)$, and $(x_n, b)$, we conclude that for sufficiently large $k$ $y_k(x) - u(x)$ has a zero in both a right and left neighborhood of $x_1$ and in disjoint neighborhoods of $x_2, x_3, \ldots, x_n$. By Hypothesis (D), this implies $y_k(x) \equiv u(x)$ for all sufficiently large $k$. From this contradiction we conclude that $S$ is closed and the proof of the Theorem is complete.

Hartman in [29] has proven that, if (4.1) satisfies (A), (B), (C) and (D) and if all $n$-point boundary value problems do have solutions, then all $k$-point boundary value problems for (4.1) also have unique solutions for $2 \leqslant k \leqslant n - 1$. Thus, if all five Hypotheses (A), (B), (C), (D) and (E) are satisfied by (4.1), then all

$k$-point boundary value problems for $2 \leqslant k \leqslant n$ do have solutions which are unique.

The proof of Theorem 4.5 for the case $n = 3$ given in [25] involved different arguments which did not require the assumption that initial value problems have unique solutions. Klaasen [30] has proven a version of Theorem 4.3 in which it is not assumed that initial value problems have unique solutions and, in an as yet unpublished paper, has used this result to prove Theorem 4.5 without assuming the uniqueness of solutions of initial value problems.

We conclude with some remarks about the compactness Hypothesis (E) of Theorem 4.5. It is an open question as to whether it is actually implied by the Hypotheses (A), (B) and (D). In [25] it is shown that this is the case for $n = 3$ and, from the proof of Theorem 4.2 it follows that (A) and (B) alone suffice in the case $n = 2$. It is not true that (A) and (B) alone imply (E) or even that (A), (B) and (C) imply (E) for $n \geqslant 3$. In fact the sequence of solutions $\{y_k(x)\}_{k=1}^{\infty}$ of the initial value problems

$$y''' + (y')^3 = 0, y(0) = y'(0) = 0, \qquad y''(0) = k,$$

furnishes a counterexample for $n = 3$. The question of whether or not (A), (B) and (D) imply (E) when $n \geqslant 4$ remains open.

An account of much of the recent work in nonlinear boundary value problems may be found in [31].

## REFERENCES


1. P. Hartman, *Ordinary Differential Equations*, Wiley, New York, 1964.
2. J. T. Schwartz, *Nonlinear Functional Analysis*, Gordon and Breach, New York, 1969.
3. E. A. Coddington and N. Levinson, *Theory of Ordinary Differential Equations*, McGraw-Hill, New York, 1955.
4. D. V. V. Wend, "On the zeros of solutions of some linear complex differential equations," *Pacific J. Math.*, **10** (1960), 713–722.
5. P. R. Beesack, "On the Green's function of an $N$-point boundary value problem," *Pacific J. Math.*, **12** (1962), 801–812.

6. Z. Nehari, "On an inequality of P. R. Beesack," *Pacific J. Math.*, **14** (1964), 261–263.

7. K. M. Das and A. S. Vatsala, "On the Green's function of an $N$-point boundary value problem," *Trans. Amer. Math. Soc.*, **182** (1973), 469–480.

8. G. B. Gustafson, "A Green's function convergence principle with applications to computation and norm estimates," *Rocky Mountain J. Math.*, **6** (1976), 457–492.

9. K. W. Schrader, "Existence theorems for second order boundary value problems," *J. Differential Equations*, **5** (1969), 572–584.

10. G. Klaasen, "Differential inequalities and existence theorems for second and third order boundary value problems," *J. Differential Equations* **10** (1971), 529–537.

11. M. Nagumo, "Uber die Differentialgleichung $y'' = f(x, y, y')$," *Proc. Phys.-Math. Soc. Japan, Ser. 3*, **19** (1937), 861–866.

12. L. K. Jackson, "Subfunctions and second order ordinary differential inequalities," *Advances in Math.*, **2** (1968), 307–363.

13. L. Erbe, "Nonlinear boundary value problems for second order differential equations," *J. Differential Equations*, **7** (1970), 459–472.

14. S. R. Bernfeld, V. Lakshmikantham, and S. Leela, "Nonlinear boundary value problems and several Lyapunov functions," *J. Math. Anal. Appl.*, **42** (1973), 545–553.

15. J. W. Bebernes and R. Wilhelmsen, "A general boundary value problem technique," *J. Differential Equations,* **8** (1970), 404–415.

16. L. K. Jackson and G. Klaasen, "A variation of the topological method of Wazewski," *SIAM J. Appl. Math.*, **20** (1971), 124–130.

17. J. Kaplan, A. Lasota, and J. Yorke, "An application of the Wazewski retract method to boundary value problems," *Zeszyty Nauk. Uniw. Jagiello. Prace Mat.*, **16** (1974), 7–14.

18. W. G. Kelley, "Some existence theorems for $n$th order boundary value problems," *J. Differential Equations*, **18** (1975), 158–169.

19. L. K. Jackson, "A Nagumo condition for ordinary differential equations," *Proc. Amer. Math. Soc.,* **57** (1976), 93–96.

20. A. Lasota and Z. Opial, "On the existence and uniqueness of solutions of a boundary value problem for an ordinary second order differential equation," *Colloq. Math.*, **18** (1967), 1–5.

21. K. Schrader and P. Waltman, "An existence theorem for nonlinear boundary value problems," *Proc. Amer. Math. Soc.*, **21** (1969), 653–656.

22. L. F. Shampine, "Existence and uniqueness for nonlinear boundary value problems," *J. Differential Equations*, **5** (1969), 346–351.

23. P. Waltman, "Existence and uniqueness of solutions of boundary value prob-

lems for two-dimensional systems of nonlinear differential equations," *Trans. Amer. Math. Soc.*, **153** (1971), 223–234.

24. Shui-Nee Chow and A. Lasota, "On boundary value problems for ordinary differential equations," *J. Differential Equations*, **14** (1973), 326–337.
25. L. K. Jackson and K. W. Schrader, "Existence and uniqueness of solutions of boundary value problems for third order differential equations," *J. Differential Equations,* **9** (1971), 46–54.
26. P. Hartman, "On *n*-parameter families and interpolation problems for nonlinear ordinary differential equations," *Trans. Amer. Math. Soc.*, **154** (1971), 201–226.
27. G. Klaasen, "Existence theorems for boundary value problems for *n*th order ordinary differential equations," *Rocky Mountain J. Math.*, **3** (1973), 457–473.
28. E. H. Spanier, *Algebraic Topology*, McGraw-Hill, New York, 1966.
29. P. Hartman, "Unrestricted *n*-parameter families," *Rend. Circ. Mat. Palermo*, (2) **7** (1958), 123–142.
30. G. Klaasen, "Continuous dependence for *N*-point boundary value problems," *SIAM J. Appl. Math.*, **29** (1975), 99–102.
31. S. R. Bernfeld and V. Lakshmikantham, *An Introduction to Nonlinear Boundary Value Problems*, Academic Press, New York, 1974.

# FUNCTIONAL ANALYSIS AND BOUNDARY VALUE PROBLEMS

*Jean Mawhin*

Boundary value problems for differential equations have been of extreme importance in the development of various parts of mathematics and of its applications to natural sciences. The part played by the theory of linear boundary value problems for linear ordinary differential equations in the creation and growth of functional analysis has been sufficiently emphasized to be reproduced here. For nonlinear differential equations, it was the periodic boundary value problem which led Poincaré to his deep researches in the qualitative theory of differential equations from which the modern theory of dynamical systems as well as important parts of algebraic and differential topology have emerged. Nonlinear boundary value problems for ordinary and partial differential equations were also the basic motivation for the extension, by Birkhoff, Kellogg, Schauder, Leray, Caccioppoli, and Rothe, of the fixed point theory to infinite dimensional vector spaces, and for the creation of various branches of nonlinear functional analysis like the variational methods, the theory of monotone operators, the projection methods, etc. Besides the more or less classical boundary value problems occurring in mechanics and physics, the modern invasion of mathematics into chemistry, biology, economy and behavior sciences constantly furnish to the curiosity of mathematicians new types of boundary value problems, e.g., in the theory of chemical reactions and ecology problems, which either illustrate known approaches or suggest new ones. In what follows, we shall try to describe, in technically simple situations, some aspects of this relation between functional analysis and boundary value problems.

## 1. BOUNDARY VS INITIAL VALUE PROBLEMS FOR ORDINARY DIFFERENTIAL EQUATIONS

If

$$f : [a, b] \times \mathbf{R}^n \to \mathbf{R}^n, \quad (t, x) \mapsto f(t, x)$$

is a given function and $S$ is a given set in $\mathbf{R}^n \times \mathbf{R}^n$, a *boundary value problem* (shortly BVP) for the vector ordinary differential equation

$$x' = f(t, x) \tag{1.1}$$

where $x' = dx/dt$, is the search of a solution $x$ of equation (1.1) which is defined upon $[a, b]$ and such that

$$(x(a), x(b)) \in S. \tag{1.2}$$

Condition (2) is called the *boundary condition* (shortly BC) because it contains only the values of the solution at the boundary points $a$ and $b$ of $[a, b]$. An important example of BC is given by the *periodic* BC

$$x(a) - x(b) = 0 \tag{1.3}$$

for which $S$ is the vector subspace of $\mathbf{R}^n \times \mathbf{R}^n$ given by

$$S = \{(y, z) : y - z = 0\}.$$

More generally one can consider linear homogeneous BC

$$Ax(a) - Bx(b) = 0 \tag{1.4}$$

where $A$ and $B$ are constant $(n \times n)$-matrices. For the vector second order differential equation

$$y'' = g(t, y, y') \tag{1.5}$$

with $g : [a, b] \times \mathbf{R}^m \times \mathbf{R}^m \to \mathbf{R}^m$, written in the form (1.1) by letting $n = 2m$, $x = (y, y')$,

$$f(t, x) = (x_{m+1}, \ldots, x_n, g_1(t, x_1, \ldots, x_n), \ldots, g_m(t, x_1, \ldots, x_n)),$$

special choices of $A$ and $B$ lead to BC which are of particular importance in applications. Let us quote the *Picard or Dirichlet* BC

$$y(a) = y(b) = 0,$$

$$\left(A = \operatorname{diag}(I_m, 0_m), \quad B = \begin{pmatrix} 0_m & 0_m \\ -I_m & 0_m \end{pmatrix}\right),$$

with $I_m$ (resp. $0_m$) the identity (resp. zero) $(m \times m)$ matrix, the *Neumann* BC

$$y'(a) = y'(b) = 0,$$

$$\left(A = \begin{pmatrix} 0_m & I_m \\ 0_m & 0_m \end{pmatrix}, B = \operatorname{diag}(0_m, -I_m)\right),$$

the *mixed* BC

$$y(a) - y'(a) = y(b) - y'(b) = 0,$$

$$\left(A = \begin{pmatrix} I_m & -I_m \\ 0_m & 0_m \end{pmatrix}, B = \begin{pmatrix} 0_m & 0_m \\ -I_m & I_m \end{pmatrix}\right),$$

and the *periodic* BC

$$y(a) - y(b) = 0, \quad y'(a) - y'(b) = 0,$$

$(A = I_{2m}, B = I_{2m})$.

Of course, *nonlinear* BC (i.e., BC for which $S$ is not affine) can also be imposed, as for example, in the case of (1.5),

$$h_1(y(a), y'(a)) = h_2(y(b), y'(b)) = 0,$$

with $h_i : \mathbf{R}^m \times \mathbf{R}^m \to \mathbf{R}^m$, $i = 1, 2$.

BVP are fundamentally different from the more familiar *initial value* or *Cauchy problem* (shortly IVP) for equation (1.1) which, given $t_0 \in [a, b]$ and $x_0 \in \mathbf{R}^n$, consists in finding a neighborhood $N$ of $t_0$ in $[a, b]$ and a solution of (1.1) defined upon $N$ and such that

$$x(t_0) = x_0.$$

Thus, the IVP is a *local* problem in that the asked solution is not required to exist upon a given interval but only in a neighborhood of the initial value $t_0$. BVP are *global* problems because the solution must exist upon a given interval and verify some relations at the boundary points of this interval. If (1.5) with $m = 1$ describes the motion of a unit mass in the field of forces described by the function $g$, solving an IVP for (1.5) answers the question: given an instant $t_0$, an initial position $y_0$ and an initial velocity $v_0$ at $t_0$, what will be the motion of the unit mass in a sufficiently near future? A BVP for the same model would answer the question: given positions $y_0$ and $y_1$, does there exist, in the field of forces described by $g$, a motion which carries the unit mass from $y_0$ at the instant $a$ to $y_1$ at the instant $b$? Because everybody expects and has observed that, given an initial position and an initial velocity, a material point has some motion, the conformity of the mathematical equation to the mechanical problem requires a fairly general positive answer to the existence problem for the IVP. Since Cauchy we know that it is effectively the case when $g$ only satisfies *regularity* conditions generally satisfied by equations describing natural phenomena. For a BVP the existence question can receive a positive or a negative answer whatever the smoothness of the equation and depends essentially in our mechanical example upon the geometry of the field of forces and of boundary conditions. We can therefore expect, and it is the case, that existence theorems for BVP will require specific assumptions of qualitative and quantitative nature upon the differential equation and the BC. The above discussion also makes natural the fact that IVP and BVP can be quite different with respect to the uniqueness of their solutions. Uniqueness in the IVP can be expected, and is proved, for sufficiently regular differential equations although many BVP have several solutions whatever the smoothness of the equation.

## 2. SHOOTING METHODS VS FUNCTIONAL ANALYTIC APPROACH FOR BOUNDARY VALUE PROBLEMS

Poincaré [32] introduced, in the case of periodic BC, a method of resolution of BVP which makes use of a preliminary knowledge of the IVP for the studied differential equation. Let us assume that equation (1.1) has, for each $u \in \mathbf{R}^n$, a unique solution $x(.; a, u)$ such that

$$x(a; a, u) = u \tag{2.1}$$

and which is defined upon $[a, b]$. Then condition (1.2) becomes

$$(u, x(b; a, u)) \in S \tag{2.2}$$

and hence the BVP (1.1)–(1.2) is reduced to finding the $u \in \mathbf{R}^n$ which verify (2.2). In the usual case where $S$ is defined by equating to zero a given function of $x(a)$ and $x(b)$, one has to solve an equation in the unknown $u$ which is in general nonlinear and moreover not explicitly determined, because it is already the case for $x(b; a, u)$. Another difficulty comes from the verification of the existence over $[a, b]$ of the solutions of the IVP, which is already a global problem for equation (1.1).

In the case of BC (1.4), one has to solve the equation in $\mathbf{R}^n$

$$Au - Bx(b; a, u) = 0 \tag{2.3}$$

which, in the periodic case, is equivalent to the determination of the fixed points of the mapping

$$u \mapsto x(b; a, u)$$

usually referred as the *Poincaré–Andronov* or *translation operator*. In the case of the BVP (1.5) with Picard BC, the above approach specializes, under the name of *shooting method*, as follows: Let $y(.; u)$ be the solution of (1.5) such that

$$y(a; u) = 0, \quad y'(a; u) = u.$$

One then tries to choose $u \in \mathbf{R}^m$ in such a way that

$$y(b; u) = 0,$$

and the corresponding solution clearly satisfies the Picard BC. More sophisticated techniques, like for example the Wazewsky's method [3], also make use of the corresponding IVP in an essential way.

A more recent and more direct approach, the one we shall emphasize in this paper, is in the spirit of functional analysis. It consists basically in considering the problem (1.1)–(1.2) as an abstract equation in a suitable space $X$ of mappings from $[a, b]$ into $\mathbf{R}^n$ chosen in such a way that the solutions of the abstract equation in $X$ are the solutions of the BVP (1.1)–(1.2). Such an approach was suggested by Birkhoff and Kellogg [4], together with their extension of the Brouwer fixed point theorem to some function spaces, and can be considered as the abstract setting of the reduction of the problem to an integral equation, as for example in Picard's earlier work [31]. Although requiring usually more sophisticated mathematical tools, the functional analytic approach has several advantages:

1. No preliminary study of the corresponding IVP is required, which weakens the needed regularity assumptions and makes possible extensions to partial differential equations.

2. Tedious calculations are usually minimized and the algebraic and topological structures of the problem are much more striking.

3. Algorithms for the approximate determination of solutions can often be naturally introduced.

In the following sections we shall try, using simple but characteristic examples, to illustrate those facts. We shall begin by the more simple case of problems which are perturbations of equations for which the solutions of the considered BVP are known. Later we shall go to more general cases.

## 3. BVP FOR SMALL PERTURBATIONS OF LINEAR DIFFERENTIAL EQUATIONS: A FIRST EXAMPLE

We shall be interested in this section in the vector differential equation

$$x' = \varepsilon f(t, x) \tag{3.1}$$

where $f : I \times \mathbf{R}^n \to \mathbf{R}^n$ is continuous and such that $\partial f / \partial x_i$ ($i = 1, \ldots, n$) exist and are continuous on $I \times R^n$, $I = [0, 1]$ and $\varepsilon \in \mathbf{R}$. We shall study the very simple BVP for (3.1) consisting in the following global IVP: find a solution of (3.1) defined on $I$ and satisfying the BC

$$x(0) = 0. \tag{3.2}$$

It is easily checked that problem (3.1)–(3.2) is equivalent to finding a continuous mapping $x : I \to \mathbf{R}^n$ such that, for each $t \in I$,

$$x(t) = \varepsilon \int_0^t f(s, x(s))\, ds. \tag{3.3}$$

If $X$ denotes the (Banach) space of continuous mappings $x : I \to \mathbf{R}^n$ such that $x(0) = 0$ with the norm

$$|x| = \max_{t \in I} |x(t)|$$

where $|x(t)|$ denotes some fixed norm of the element $x(t)$ of $\mathbf{R}^n$, and if we define on $X$ the mapping $N$ by

$$(Nx)(t) = \int_0^t f(s, x(s))\, ds, \quad t \in I \tag{3.4}$$

it is not hard to show that $N$ maps $X$ into itself. Moreover, if $x, y \in B[R]$, the closed ball of center $O$ and radius $R$ in $X$, we have, using the mean value theorem,

$$|(Nx)(t) - (Ny)(t)| \leq \left| \int_0^t (f(s, x(s)) - f(s, y(s)))\, ds \right|$$
$$\leq \sup_{\substack{t \in I \\ |z| \leq R}} |f'_x(t, z)| |x - y|,$$

where $f'_x(t, z)$ denotes the $(n \times n)$ matrix with components

$$\frac{\partial f_i}{\partial x_j}(t, z) \quad (i, j = 1, \ldots, n).$$

Therefore, we can find $k(R) > 0$ such that, for $x, y \in B[R]$,

$$|Nx - Ny| \leq k(R)|x - y|.$$

Thus, equation (3.3) is equivalent to the equation in $X$

$$x = \varepsilon N x, \tag{3.5}$$

i.e., to finding the fixed points of $\varepsilon N$, with $\varepsilon N$ such that, for $x, y \in B[R]$,

$$|\varepsilon N x - \varepsilon N y| \leq \varepsilon k(R)|x - y|. \tag{3.6}$$

But, for mappings in Banach (and even complete metric) spaces verifying conditions of type (3.6) one has the simple but basic *Banach fixed point theorem* or *contraction mapping theorem* proved, e.g., in Graves [12].

THEOREM 3.1: *Let $X$ be a Banach space and $F : B[R] \to X$ be* *such that there exists* $k \in [0, 1[$ *for which, when* $x, y \in B[R]$,

$$|Fx - Fy| \leqslant k|x - y| \tag{3.7}$$

*and*

$$|F(0)| \leqslant (1 - k)R. \tag{3.8}$$

*Then there exists in* $B[R]$ *a unique* $x$ *such that*

$$x = Fx,$$

*(i.e., a unique fixed point of* $F$ *in* $B[R]$*) and*

$$x = \lim_{n \to \infty} x_n,$$

*with* $x_0 \in B[R]$ *arbitrary and*

$$x_{n+1} = Fx_n, \; n = 0, 1, \ldots .$$

*If* $F : X \to X$ *and satisfies* (3.7) *for all* $x, y \in X$, *the same conclusions hold.*

We can apply this theorem to equation (3.5) by taking $F = \varepsilon N$ and noting that $R > 0$ being given and $\varepsilon_0$ being defined by

$$\varepsilon_0 = \left[ R^{-1}|N(0)| + k(R) \right]^{-1},$$

we shall have, for each $\varepsilon \in ]-\varepsilon_0, \varepsilon_0[$,

$$\varepsilon k(R) < 1, \quad |\varepsilon N(0)| \leqslant (1 - \varepsilon k(R))R,$$

which are conditions (3.7) and (3.8) for $\varepsilon N$. Hence applying Theorem 3.1 to (3.5) and translating the results in terms of differential equations, we obtain the following

THEOREM 3.2: *Let* $f$ *satisfying the regularity conditions stated above. Then, for each* $R > 0$, *there exists* $\varepsilon_0 > 0$ *such that, for every*

$$\varepsilon \in ]-\varepsilon_0, \varepsilon_0[,$$

*the* BVP (3.1)–(3.2) *has a unique solution* $x^\varepsilon$ *verifying*

$$|x^\varepsilon(t)| \leqslant R, \quad t \in I.$$

*Moreover,* $x^\varepsilon$ *can be obtained as the (uniform) limit of successive approximations* $x_n^\varepsilon$ *with (say)* $x_0^\varepsilon = 0$ *and*

$$x_{n+1}^\varepsilon(t) = \varepsilon \int_0^1 f(s, x_n^\varepsilon(s))\, ds, \quad n = 0, 1, \dots .$$

It is also easily checked that

$$x^\varepsilon \to 0$$

as $\varepsilon \to 0$ uniformly in $t \in I$.

## 4. BVP FOR SMALL PERTURBATIONS OF LINEAR DIFFERENTIAL EQUATIONS: A SECOND EXAMPLE

Let us consider now, for equation (3.1) with the regularity assumptions of section 3, the periodic BVP

$$x(0) - x(1) = 0. \tag{4.1}$$

It is easy to show that finding a solution of the BVP (3.1)–(4.1) is equivalent to finding a continuous mapping $x : I \to \mathbf{R}^n$ such that, for $t \in I$,

$$x(t) - x(0) = \varepsilon \int_0^t f(s, x(s))\, ds \tag{4.2}$$

and which satisfies (4.1). Let now $X$ be the (Banach) space of continuous mappings $x : I \to \mathbf{R}^n$ verifying (4.1), with the norm

$$|x|_X = \max_{t \in I} |x(t)|,$$

and define $N$ by relation (3.4). Contrarily to the previous example,

$N$ does not in general map $X$ into itself, because usually,

$$(Nx)(1) = \int_0^1 f(s, x(s))\, ds \neq 0 = (Nx)(0).$$

In fact, one has, for $x \in X$, $t \in I$,

$$\int_0^t f(s, x(s))\, ds = \int_0^t \left[ f(s, x(s)) - \int_0^1 f(u, x(u))\, du \right] ds$$

$$+ t \int_0^1 f(s, x(s))\, ds, \tag{4.3}$$

and hence $(Nx)(t)$ is of the form $z(t) + tc$ with $z \in X$ such that

$$z(0) = z(1) = 0$$

and $c \in \mathbf{R}^n$. Thus, denoting by $Y$ the (Banach) space of mappings $y : I \to \mathbf{R}^n$ which are of the form

$$y(t) = tc + z(t), \quad t \in I,$$

with $c \in \mathbf{R}^n$ and

$$z \in \{z \in X : z(0) = 0\},$$

with the norm

$$|y|_Y = |c| + |z|_X,$$

we see that $N$ maps $X$ into $Y$. Also, if we define on $X$ the linear mapping $L$ by

$$(Lx)(t) = x(t) - x(0), \quad t \in I, \tag{4.4}$$

we check at once that $L$ maps $X$ into $Y$ (because $x(.) - x(0)$ is in $X$ and $(Lx)(0) = 0$) and is bounded. Therefore, solving (4.2)–(4.1) is equivalent to finding a solution in $X$ of the abstract equation

$$G(x, \varepsilon) \equiv Lx - \varepsilon Nx = 0 \tag{4.5}$$

where $G : X \times \mathbf{R} \to Y$.

From our assumptions upon $f$, it is not too hard to check that $G$ is (Frechet) differentiable with respect to $(x, \varepsilon)$ and hence we can try to solve (4.5) using the implicit function theorem in Banach spaces as given, e.g., in Graves [12].

THEOREM 4.1: *$X$, $Y$ and $Z$ being Banach spaces, let*

$$G : X \times Z \to Y$$

*be continuously (Frechet) differentiable on $X \times Z$. Suppose that there exists $(x_0, z_0) \in X \times Z$ such that*

$$G(x_0, z_0) = 0$$

*and such that the (Frechet) partial derivative $G'_x(x_0, z_0)$ is a linear homeomorphism from $X$ onto $Y$. Then there exist positive constants $a$, $b$ and a continuous mapping*

$$\xi : B[b] \subset Z \to B[a] \subset X$$

*such that*

(i) $G(\xi(z), z) = 0$ *for each* $z \in B[b]$;
(ii) *if* $G(x, z) = 0$, $x \in B[a]$, $z \in B[b]$, *then* $x = \xi(z)$.

Thus (i) is an existence conclusion and (ii) a local uniqueness assertion.

Let us try to apply Theorem 4.1 to equation (4.5) with $X$, $Y$ defined above, $Z = \mathbf{R}$, $z_0 = 0$. Clearly each constant mapping $x_0 : t \mapsto x_0$ verifies

$$G(x_0, 0) = Lx_0 = x_0 - x_0 = 0,$$

but, as is easily computed,

$$G'_x(x_0, 0) = Lx_0 = 0$$

cannot be a homeomorphism and hence Theorem 4.1 is not applicable. The reason is that we have not really used the whole information contained in the special structure of equation (4.5). Let us first note that if problem (4.1)–(4.2) has a solution $x^\varepsilon$ for sufficiently small $\varepsilon$, then necessarily

$$\int_0^1 f(s, x^\varepsilon(s))\, ds = 0,$$

and hence, if $x^\varepsilon$ has a limit $x_0$ when $\varepsilon \to 0$, one has, by continuity,

$$\int_0^1 f(s, x_0)\, ds = 0,$$

a condition which must therefore appear in some form in each sufficient condition for the existence of a solution. Now coming back to (4.5) we see at once that the range Im $L$ of $L$ is the proper subspace of $Y$

$$Y_1 = \{y \in Y : y \in X \quad \text{and} \quad y(0) = 0\}.$$

Also (4.3) can be written

$$(Nx)(t) = (Nx)(t) - t(Nx)(1) + t(Nx)(1), \quad t \in I,$$

i.e., denoting by $Q$ the projector along $Y_1$ onto its direct summand

$$Y_2 = \{y \in Y : y(t) = tc, c \in \mathbf{R}^n, t \in I\},$$

in $Y$,

$$Nx = (I - Q)Nx + QNx. \tag{4.6}$$

Clearly, $Q$ is explicitly defined by

$$(Qy)(t) = ty(1), \quad t \in I,$$

and is a linear bounded operator in $Y$. Now equation (4.5) can be written equivalently

$$Lx - \varepsilon(I - Q)Nx - \varepsilon QNx = 0,$$

i.e.,

$$G_1(x, \varepsilon) + \varepsilon G_2 x = 0,$$

where $G_1(x, \varepsilon) = Lx - \varepsilon(I - Q)Nx$ and $\varepsilon G_2 x = -\varepsilon QNx$, which show that $G_1$ and $G_2$ respectively take their values in the complementary vector subspaces $Y_1$ and $Y_2$ of $Y$. Therefore, for each $\varepsilon \neq 0$, equation (4.5) is surely equivalent to

$$H(x, \varepsilon) \equiv Lx - \varepsilon(I - Q)Nx - QNx = 0 \tag{4.7}$$

which, when $\varepsilon = 0$, reduces to

$$Lx - QNx = 0,$$

or, $L$ and $Q$ taking their values in complementary subspaces of $Y$,

$$Lx = 0, \quad QNx = 0. \tag{4.8}$$

But we have already observed that the null-space of $L$ is given by

$$\ker L = \{x \in X : x \text{ is a constant mapping}\}$$

and hence (4.8) is equivalent to solving equation

$$QNx = 0$$

in $\ker L$, i.e., explicitly, identifying naturally the constant mappings from $I$ into $\mathbf{R}^n$ and $\mathbf{R}^n$ and noting that $tc = 0$ for $t \in I$ if and only if $c = 0$,

$$F(a) \equiv \int_0^1 f(s, a)\, ds = 0, \tag{4.9}$$

with $a \in \mathbf{R}^n$, which essentially is the necessary condition obtained above.

In order to apply Theorem 4.1 to the new equation (4.7), let $a^*$ be a zero of $F$, which means that the constant mapping $a^* : t \mapsto a^*$ verifies

$$H(a^*, 0) = 0.$$

An easy computation shows that, if $u \in X$,

$$(H_x'(a^*, 0)u)(t) = u(t) - u(0) - t\int_0^1 f_x'(s, a^*)u(s)\, ds,$$

and to find conditions under which $H'_x(a^*, 0)$ is a homeomorphism of $X$ onto $Y$, we have to study the equation

$$u(t) - u(0) - t\int_0^1 f'_x(s, a^*)u(s)\,ds = v(t), \quad t \in I, \tag{4.10}$$

with $v \in Y$. Letting

$$v(t) = w(t) + tc, \quad u(t) = u(0) + h(t),$$

where $w \in Y_1$, $c = v(1) \in \mathbf{R}^n$, $h(t) = u(t) - u(0)$, (4.10) can be written

$$h(t) - t\left[\int_0^1 f'_x(s, a^*)\,ds\right]u(0) - t\int_0^1 f'_x(s, a^*)h(s)\,ds = tc + w(t),$$

which is equivalent to the system

$$-\left[\int_0^1 f'_x(s, a^*)\,ds\right]u(0) - \int_0^1 f'_x(s, a^*)h(s)\,ds = c,$$

$$h(t) = w(t), \quad t \in I.$$

Therefore (4.10) will have the unique solution

$$\begin{aligned} u(t) &= -\left[F'(a^*)\right]^{-1}\left[c + \int_0^1 f'_x(s, a^*)w(s)\,ds\right] + w(t) \\ &= -\left[F'(a^*)\right]^{-1}\left[v(1) + \int_0^1 f'_x(s, a^*)(v(s) - sv(1))\,ds\right] \\ &\quad + v(t) - tv(1), \quad t \in I, \end{aligned} \tag{4.11}$$

if we assume that

$$F'(a^*) = \int_0^1 f'_x(s, a^*)\,ds$$

is an automorphism of $\mathbf{R}^n$, i.e., if we assume that

$$\det F'(a^*) \neq 0. \tag{4.12}$$

It is easily checked using (4.11) that the linear mapping $H_x'(a^*, 0)^{-1} : Y \to X, w \mapsto u$ is bounded and hence that $H_x'(a^*, 0)$ is a linear homeomorphism of $X$ onto $Y$. Thus applying Theorem 4.1 to equation (4.7) and translating the results in terms of differential equations, we obtain the following

THEOREM 4.2: *Let $f$ satisfying the regularity conditions stated above, and assume that equation* (4.9) *has a solution $a^*$ such that* (4.12) *holds. Then there exist $r > 0$ and $\varepsilon_0 > 0$ such that, for each $\varepsilon \in [-\varepsilon_0, \varepsilon_0]$, the* BVP (3.1)–(4.1) *has a unique solution $x^\varepsilon$ verifying*

$$|x^\varepsilon(t) - a^*| \leqslant r$$

*for $t \in I$.*

It is also easily checked that

$$x^\varepsilon(t) \to a^*$$

as $\varepsilon \to 0$, uniformly in $t \in I$, and it follows also from the proof of the implicit function theorem that $x^\varepsilon$ can be obtained as the (uniform) limit of successive approximations $x_n^\varepsilon$ defined in a rather lengthy way that we shall not describe here.

## 5. BVP FOR SMALL PERTURBATIONS OF LINEAR DIFFERENTIAL EQUATIONS: NON-CRITICAL VS CRITICAL CASES

In this section we shall compare and discuss the assertions of Theorems 3.2 and 4.2. In Theorem 3.2 one requires only restrictions upon the smoothness and the size of the function $\varepsilon f$ to get existence and uniqueness of the BVP, and the conditions are always satisfied if $f$ depends only upon $t$. Theorem 3.2 essentially says that the BVP (3.1)–(3.2), which has a unique solution ($x = 0$) when $\varepsilon = 0$, conserves this property for sufficiently small values of

$|\varepsilon|$, i.e., *regardless of the nonlinear perturbation*. Also, when $\varepsilon \to 0$, the solution of the perturbed problem approaches the solution of the unperturbed problem. Thus, for this BVP, the cases corresponding to $\varepsilon = 0$ and to $\varepsilon \neq 0$ but small in absolute value are neither qualitatively nor quantitatively very different. The situation is quite different in Theorem 4.2, where the existence of a solution not only depends upon some smoothness of $f$ and some smallness of $|\varepsilon|$, but also of the existence of a solution for equation (4.9) in which the nonlinear perturbation appears in a basic way. Thus the qualitative behavior of the nonlinear term is crucial for the existence or the non-existence of a solution, which is not so surprising if we note that, in our existence result, perturbations of the type $\varepsilon h(t)$ with

$$\int_0^1 h(s)\, ds \neq 0,$$

and for which the considered periodic BVP has clearly no solutions, must be ruled out, as well as $f(t, x) \equiv 0$ for which uniqueness fails. Thus, in Theorem 4.2 we can say that we have existence and (local) uniqueness *because of the nonlinear perturbation* and it necessarily plays a basic role in the assertion. Moreover, if equation (4.9) has several solutions verifying (4.12), our BVP has, for all sufficiently small values of $|\varepsilon|$, several solutions which, when $\varepsilon \to 0$, approach *some* solutions of the corresponding BVP with $\varepsilon = 0$ (here constant functions) which are selected from the set of solutions when $\varepsilon = 0$ by the nonlinear perturbation. Hence the qualitative picture of the solutions for $\varepsilon = 0$ (here an $n$-dimensional subspace of $X$) can dramatically change for $\varepsilon \neq 0$, leading often to a "crystallization" of the infinite family of solutions when $\varepsilon = 0$ to a finite one when $\varepsilon \neq 0$. Such a qualitative change in the nature of the solutions for some value of the parameter is generally referred as a *bifurcation* or a *branching* of the solution and, for this reason, equation (4.9) is called the *bifurcation* or *branching equation* and it always appears in one form or another whatever the method used to solve the problem. Poincaré, Lyapunov, and E. Schmidt were the pioneers in the study of bifurcation phenomena in nonlinear differential and integral equations.

The fundamental reason for the profound difference in treatment and results for the two considered examples can be found in the different properties of their respective linear parts, i.e., of the corresponding problems for $\varepsilon = 0$. In the case of (3.1)–(3.2), we obtain, if $\varepsilon = 0$,

$$x' = 0, \quad x(0) = 0, \tag{5.1}$$

which obviously admits the unique solution $x = 0$, and the corresponding linear non-homogeneous problem

$$x' = h(t), \quad x(0) = 0,$$

with $h : I \to \mathbf{R}^n$ continuous, has the unique solutions

$$x(t) = \int_0^t h(s)\, ds, \quad t \in I.$$

Therefore, in problem (3.1)–(3.2) we perturb slightly a linear one-to-one and onto mapping and prove that the complete equation still has a (locally) unique solution. A linear differential equation having the properties of $x'$ with respect to the BC (3.2) is said to be *non-critical* or *non-resonant with respect to the* BC and the approach described in section 3 can be, modulo greatest technical difficulties only, extended to more general BVP of the type

$$x' = C(t)x + \varepsilon f(t, x), \tag{5.2}$$

$$Ax(0) - Bx(1) = 0, \tag{5.3}$$

where the constant $(n \times n)$ matrices $A$, $B$, with the $(n \times 2n)$ matrix $(A, B)$ of rank $n$, and the continuous $(n \times n)$ matrix function $C$ are such that the associated linear BVP

$$\begin{aligned} x' &= C(t)x \\ Ax(0) - Bx(1) &= 0 \end{aligned} \tag{5.4}$$

is *non-critical*, i.e., only admits the trivial solution. In this case, one can introduce a matrix function

$$(t, s) \mapsto G(t, s)$$

piecewise continuous upon $I \times I$, the *Green matrix associated to* (5.4) which has the property that the unique solution of the linear non-homogeneous BVP

$$x' = C(t) + h(t)$$

$$Ax(0) - Bx(1) = 0$$

is given by

$$x(t) = \int_0^1 G(t, s)h(s)\, ds, \quad t \in I.$$

This Green matrix is used to reduce the BVP (5.2)–(5.3) to the equivalent integral equation

$$x(t) = \int_0^1 G(t, s)f(s, x(s))\, ds$$

which corresponds to (3.3) in our more general case. In the example considered in section 3, the associated Green matrix is given by

$$G(t, s) = \begin{cases} 1 & \text{if } 0 \leqslant s \leqslant t < 1, \\ 0 & \text{if } 0 \leqslant t < s \leqslant 1. \end{cases}$$

When $\varepsilon = 0$, the BVP (3.1)–(4.1) reduces to

$$x' = 0, \quad x(0) - x(1) = 0$$

which admits the $n$-dimensional vector space of solutions

$$x(t) = c, \quad c \in \mathbf{R}^n, \quad t \in I,$$

and the corresponding linear non-homogeneous problem

$$x' = h(t), \quad x(0) - x(1) = 0$$

has a solution if and only if

$$\int_0^1 h(s)\, ds = 0.$$

Thus the abstract linear mapping corresponding to our problem with $\varepsilon = 0$ is neither one-to-one nor onto and it is therefore less surprising that additional assumptions upon the nonlinear perturbation should be required to make the complete equation solvable. The linear part of our differential equation is said in this case to be *critical* or *resonant with respect to the* BC and the approach described in section 4 can be extended with the same basic tools to more general BVP of type (5.2)–(5.3) when the corresponding linear problem (5.4) has nontrivial solutions, i.e., when the linear differential equation is *critical* with respect to the BC (5.3). In this case, the equation corresponding to (4.5) can be constructed in terms of *generalized Green matrices* that we shall not describe here.

## 6. HILBERT SPACES METHODS AND SHARP ESTIMATES IN SOME NON-CRITICAL PROBLEMS

The results stated in sections 3 and 4 hold for "sufficiently small" values of $|\varepsilon|$ and the upper value $\varepsilon_0$ for $|\varepsilon|$ furnished by those theorems is rarely the best possible one. To obtain sharp results, it is usually better, when it is possible, to formulate the problem as an abstract equation in a Hilbert space in which the norm of linear operators can often be more easily estimated. To illustrate this fact, let us consider the periodic BVP

$$\begin{gathered} x'' + g(x) = e(t), \\ x(0) - x(1) = x'(0) - x'(1) = 0, \end{gathered} \tag{6.1}$$

where $e : I \to \mathbf{R}$ is continuous, $g : \mathbf{R} \to \mathbf{R}$ of class $C^1$ and such that there exist a nonnegative integer $n$ and real numbers $p$, $q$ such that, for all $x \in \mathbf{R}$,

$$(2\pi n)^2 < p \leqslant g'(x) \leqslant q < (2\pi(n+1))^2. \tag{6.2}$$

Following the spirit of section 5, we call (6.1) a noncritical problem because in the case where $g$ is linear, conditions (6.2) imply that the linear differential equation

$$x'' + g(x) = 0$$

is noncritical with respect to the periodic BC because it admits only the trivial solution.

Let $H$ be the Hilbert space $L^2(I)$ of (Lebesgue) square-integrable real functions $x$ with the norm

$$\|x\| = \left[\int_0^1 x^2(t)\,dt\right]^{1/2}.$$

If $x \in H$, it is known that $x$ has the Fourier series

$$x(t) \sim a_0 + 2^{\frac{1}{2}} \sum_{s=1}^{\infty} (a_s \cos(2\pi st) + b_s \sin(2\pi st))$$

where

$$a_0 = \int_0^1 x(t)\,dt, \quad a_s = 2^{\frac{1}{2}} \int_0^1 x(t)\cos(2\pi st)\,dt,$$

$$b_s = 2^{\frac{1}{2}} \int_0^1 x(t)\sin(2\pi st)\,dt, \quad s = 1, 2, \ldots,$$

and the Parseval equality

$$\|x\|^2 = a_0^2 + \sum_{s=1}^{\infty} (a_s^2 + b_s^2)$$

holds. By a solution of (6.1) we shall mean here a real function $x$ which is absolutely continuous on $I$ together with $x'$ and such that $x'' \in H$. A twice continuously differentiable solution of (6.1) clearly has the above properties and, $g$ and $e$ being continuous, it follows from (6.1) that if $x$ is a solution in the sense just defined, $x''$ will be continuous and hence, for our problem, the two concepts are equivalent.

Let $\nu$ be any real number such that

$$(2\pi n)^2 < \nu < (2\pi(n+1))^2, \tag{6.3}$$

and let $h \in H$. It is easily checked, using Fourier series, that the linear differential equation

$$x'' + \nu x = h(t) \tag{6.4}$$

has a unique solution $x$, satisfying the periodic BC, given by

$$x(t) = \nu^{-1}c_0 + 2^{\frac{1}{2}} \sum_{s=1}^{\infty} \left(\nu - (2\pi s)^2\right)^{-1}(c_s \cos(2\pi st) + d_s \sin(2\pi st))$$

if

$$h(t) \sim c_0 + 2^{\frac{1}{2}} \sum_{s=1}^{\infty} (c_s \cos(2\pi st) + d_s \sin(2\pi st)).$$

Moreover,

$$\|x\|^2 = \nu^{-2}c_0^2 + \sum_{s=1}^{\infty} \left(\nu - (2\pi s)^2\right)^{-2}(c_s^2 + d_s^2)$$

$$\leqslant \left(\min\left(|\nu - (2\pi n)^2|, |\nu - (2\pi(n+1))^2|\right)\right)^{-2} \|h\|^2,$$

where we have used (6.3) and Parseval equality. Therefore, if we denote by $Ah$ the unique solution $x$ of (6.4) verifying the periodic BC, we have

$$\|Ah\| \leqslant \left(\min\left(|\nu - (2\pi n)^2|, |\nu - (2\pi(n+1))^2|\right)\right)^{-1} \|h\|, \tag{6.5}$$

and hence we have defined in this way a linear bounded operator $A : H \to H$.

Now the differential equation in (6.1) can be written equivalently

$$x'' + \nu x = \nu x - g(x) + e(t), \tag{6.6}$$

and it follows easily from (6.2) that for some $\alpha \geqslant 0$, $\beta \geqslant 0$ and all $x \in \mathbf{R}$, $t \in I$,

$$|\nu x - g(x) + e(t)| \leqslant \alpha|x| + \beta,$$

which easily implies, using Schwarz inequality, that, for each $x \in H$, the function defined a.e. on $I$ by

$$t \mapsto \nu x(t) - g(x(t)) + e(t)$$

belongs to $H$. Therefore, the mapping $N$ defined on $H$ by

$$(Nx)(t) = \nu x(t) - g(x(t)) + e(t)$$

a.e. on $I$ maps $H$ into itself, and hence equation (6.6) with the periodic BC is equivalent to the abstract equation in $H$

$$x = ANx \tag{6.7}$$

which, having a right-hand side obtained by the composition of a nonlinear and of a linear mapping, is usually called a *Hammerstein equation*.

But, using the mean-value theorem and (6.2), we obtain, for $x, y \in \mathbf{R}$,

$$\begin{aligned} |(\nu x - g(x)) - (\nu y - g(y))| &\leqslant \sup_{z \in \mathbf{R}} |g'(z) - \nu||x - y| \\ &\leqslant \max(|q - \nu|, |p - \nu|)|x - y|, \end{aligned}$$

which easily implies that, if $x, y \in H$,

$$\|Nx - Ny\| \leqslant \max(|q - \nu|, |p - \nu|)\|x - y\|,$$

and hence, using (6.5),

$$\begin{aligned} \|ANx - ANy\| &\leqslant \max(|q - \nu|, |p - \nu|) \\ &\quad \cdot \Big(\min\big(|\nu - (2\pi n)^2|, |\nu - (2\pi(n+1))^2|\big)\Big)^{-1} \\ &\quad \cdot \|x - y\| \\ &= \gamma\|x - y\|. \end{aligned} \tag{6.8}$$

A simple calculation using (6.2) and (6.3) shows now that $0 \leqslant \gamma < 1$ if and only if one takes

$$\left(q + (2\pi n)^2\right)/2 < \nu < \left(p + (2\pi(n+1))^2\right)/2$$

which is always possible. For such a value of $\nu$, we see from (6.8) that $AN$ satisfies the conditions of Theorem 3.1 and hence has a unique fixed point. Translating the result in terms of differential equations, we obtain the following

THEOREM 6.1: *If condition* (6.2) *holds, the* BVP (6.1) *has a unique solution.*

The result is sharp because, in the case of a linear function $g(x) = px$, with $p \geqslant 0$, the condition

$$(2\pi n)^2 < p < (2\pi(n+1))^2$$

is necessary and sufficient for the unique solvability of the periodic BVP (6.1) with an arbitrary continuous function $e$.

In the case of other BC for which the associated spectral problem

$$\begin{gathered} x'' - \lambda x = 0, \\ x \quad \text{satisfies the BC,} \end{gathered} \tag{6.9}$$

has a countable set of eigenvalues and a corresponding complete orthonormal set of eigenfunctions in $H$, one can use the same approach by replacing the usual Fourier series above by a development along the eigenfunctions. For example, the Picard, Neumann and mixed BVP for the differential equation in (6.1) can be treated in this way. One has only to replace in (6.2) $(2\pi n)^2$ and $(2\pi(n+1))^2$ (which are the negative of two consecutive eigenvalues of the problem (6.9) with periodic BC) by the negative of two consecutive eigenvalues of (6.9) with the considered BC.

## 7. BVP FOR STRONGLY NONLINEAR DIFFERENTIAL EQUATIONS: NON-CRITICAL CASES

The treatment of BVP which are not of the perturbational type requires more sophisticated mathematical tools, although underlying principles are still simple. As an example of a typical problem, we shall come back to the BVP of section 3, but without a small parameter in the differential equation and with less regularity conditions. Specifically, let $f : I \times \mathbf{R}^n \to \mathbf{R}^n$ be continuous and let us consider the problem of finding a solution $x$ of the equation

$$x' = f(t, x) \tag{7.1}$$

which is defined on $I$ and such that

$$x(0) = 0. \tag{7.2}$$

Using the notations of section 3, we see at once that our problem is equivalent to finding the fixed points of the mapping $N$ in $X$, with $N$ defined in (3.4). With the mere assumption of continuity for $f$, $N$ is no more Lipschitzian and hence is not a contraction. But the fact that $N$ is defined through an integration of a continuous mapping makes it a *compact* mapping in $X$, i.e., a continuous mapping transforming bounded sets into relatively compact ones. In fact, if $|x| \leqslant B$, then

$$|Nx|_X \leqslant \sup_{\substack{t \in I \\ |x| \leqslant B}} |f(t, x)| \leqslant C$$

and, if $t, t' \in I$,

$$|(Nx)(t) - (Nx)(t')| \leqslant C|t - t'|,$$

which shows that the set $\{Nx : |x| \leqslant B\}$ is equibounded and equicontinuous and hence relatively compact using Arzela–Ascoli theorem (see, e.g., Goffman [11]).

Compactness is an alternative to completeness in many existence theories and in particular the more powerful fixed point theorems for continuous mappings in $\mathbf{R}^n$ heavily depend upon the fact that those mappings take bounded closed sets into bounded closed, i.e., compact sets. Closed bounded sets being not necessarily compact in infinite-dimensional normed spaces, the extension of those fixed point techniques to such spaces requires either to work with continuous mappings on compact sets or with compact mappings on bounded sets. Of particular importance in those spaces is the following *continuation theorem of Leray and Schauder* [24].

THEOREM 7.1: *Let $X$ be a normed vector space, $\Omega \subset X$ a bounded open set and $N : \text{cl}\,\Omega \to X$ a compact mapping. If the following conditions hold:*

(a) $x \neq \lambda N x$ *for each* $x \in \text{bdry}\,\Omega$ *and each* $\lambda \in ]0, 1[$,

(b) $0 \in \Omega$,

*then, for each* $\lambda \in [0, 1]$, $\lambda N$ *has at least one fixed point in* $\text{cl}\,\Omega$.

The proof of this result is involved and will not be given here; it is interesting to note that its more difficult part consists in getting the result when $X = \mathbf{R}^n$. The underlying idea in Theorem 7.1 is the following: When $\lambda = 0$, $\lambda N$ has, by (b) the (unique) fixed point 0 in $\Omega$; if $\lambda$ is increasing from 0 to 1, fixed points of $\lambda N$ situated in $\Omega$ cannot "escape" through the boundary of $\Omega$ because of condition (a) and the art consists then in showing that the fixed points in $\Omega$ never disappear completely, when $\lambda$ varies, by coalescing. A very easy consequence of Theorem 7.1 is the famous *Schauder fixed point theorem* anticipated, for special function spaces, by Birkhoff and Kellogg [4].

THEOREM 7.2: *Let $X$ be a normed space, and $N : B[R] \to X$ a compact mapping such that $N(B[R]) \subset B[R]$. Then $N$ has at least one fixed point in $B[R]$.*

In applying Theorem 7.1 to the BVP (7.1)–(7.2), the difficult part consists obviously in finding a bounded open set $\Omega$ in $X$ such that condition (a) holds for the corresponding mapping $\lambda N$, i.e., for the associated family of BVP

$$x' = \lambda f(t, x), \quad \lambda \in ]0, 1[, \tag{7.3}$$

$$x(0) = 0. \tag{7.4}$$

Condition (a) is an *a priori* condition upon the possible solutions of (7.3)–(7.4) which has of course to be verified using the differential equation and the BC itself. This is often done by exhibiting the existence of an *a priori bound* for all possible solutions of (7.3)–(7.4), i.e., by showing that they are necessarily such that

$$|x| < R,$$

for some $R > 0$ which does not depend upon $\lambda$ and upon the solution. Then clearly the open ball $B(R)$ of center $O$ and radius $R$ verifies condition (a) of Theorem 7.1. Various techniques have been used to obtain those *a priori* estimates whose discovery can also be facilitated by the possible underlying physical interpretation of the BVP. We shall illustrate some of them for BVP (7.1)–(7.2) by simple examples.

We shall first show that some *growth restrictions upon f* can lead to *a priori* bounds for the possible solutions of (7.3)–(7.4). Assume that, for some $A \in [0, 1[$ and $C \geqslant 0$,

$$|f(t, x)| \leqslant A|x| + C \tag{7.5}$$

for all $t \in I$ and $x \in \mathbf{R}^n$. Then,

$$|Nx| = \max_{t \in I} \left| \int_0^t f(s, x(s))\, ds \right| \leqslant A|x| + C$$

and, if $x$ is a possible fixed point of $\lambda N$,

$$|x| = |\lambda N x| \leqslant A|x| + C,$$

which implies that

$$|x| \leqslant (1 - A)^{-1} C.$$

Using Theorem 7.1 with $\Omega = B((1 - A)^{-1}C)$ we immediately obtain the

THEOREM 7.3: *If condition* (7.5) *holds with* $A \in [0, 1[$ *and* $C \geqslant 0$, *then the* BVP (7.1)–(7.2) *has at least one solution.*

*A priori* estimates can also be obtained from the *geometry of the vector field defined in* $I \times \mathbf{R}^n$ *by* $f(t, x)$. To show that, let us first call an open, bounded subset $G$ of $R^n$ a *bound set relative to* (7.1) if $0 \in G$ and if, for each $x_0 \in$ bdry $G$, there exists a $C^1$ function $V_{x_0} : \mathbf{R}^n \to \mathbf{R}$ such that $V_{x_0}(x_0) = 0$ and

(i) $G \subset \{x : V_{x_0}(x) < 0\}$,

(ii) if $V'_{x_0}(x)$ denotes the value at $x$ of the gradient of $V_{x_0}$, then, for each $x_0 \in$ bdry $G$ and each $t \in I$,

$$V'_{x_0}(x_0) \cdot f(t, x_0) \neq 0, \tag{7.6}$$

where $(\cdot)$ denotes the inner product in $\mathbf{R}^n$.

In the case where (7.6) is replaced by

$$V'_{x_0}(x_0) \cdot f(t, x_0) < 0 \quad (\text{resp.} > 0) \tag{7.7}$$

$G$ will be called an *attractive* (resp. *repulsive*) *bound set*.

We can now prove an existence result for BVP (7.1)–(7.2).

THEOREM 7.4: *If there exists an attractive bound set relative to* (7.1), *then the* BVP (7.1)–(7.2) *has at least one solution such that* $x(t) \in \text{cl } G$ *for all* $t \in I$.

*Proof:* To apply Theorem 7.1, let us define the open bounded set in $X$

$$\Omega = \{x \in X : x(t) \in G \quad \text{for every } t \in I\}$$

and assume that, for some $\lambda \in ]0, 1[$, the BVP (7.1)–(7.2) has a solution $x$ such that $x \in$ bdry $\Omega$. Then, from the definition of $\Omega$, $x(t) \in \text{cl } G$ for all $t \in I$ and there exists $t_0 \in I$ such that $x(t_0) \in$ bdry $G$. As $0 \in G$ and $x(0) = 0$, one necessarily has $t_0 \in ]0, 1]$.

Suppose first that $t_0 \in ]0, 1[$ and write $x_0 = x(t_0)$. Then $V_{x_0}(x_0) = 0$ and by (i),

$$v(t) \equiv V_{x_0}(x(t)) \leqslant 0, \quad t \in I.$$

Thus, $t_0$ is an interior point of $I$ at which the $C^1$ function $v : I \to \mathbf{R}$ has a maximum, and therefore,

$$0 = v'(t_0) = V'_{x_0}(x_0) \cdot x'(t_0) = \lambda V'_{x_0}(x_0) \cdot f(t_0, x_0)$$

which contradicts (7.7). Now, if $t_0 = 1$, $v$ has in the same way a maximum at the end point 1 of $I$ and hence, if $v'(1)$ is the left derivative of $v$ at 1,

$$0 \leqslant v'(1) = V'_{x_0}(x_0) \cdot x'(1) = \lambda V'_{x_0}(x_0) \cdot f(1, x_0)$$

which still contradicts (7.7). As clearly $0 \in \Omega$, the proof is complete.

Theorem 7.4 admits several interesting special cases we shall describe now. Let $G \subset \mathbf{R}^n$ be an open bounded convex set such that $0 \in G$. Then, it is known that, for each $x_0 \in \text{bdry } G$, there exists a (not necessarily unique) $n(x_0) \in \mathbf{R}^n \setminus \{0\}$ such that $n(x_0) \cdot x_0 > 0$ for $x_0 \in \text{bdry } G$ and

$$G \subset \{x : (x - x_0) \cdot n(x_0) < 0\}. \tag{7.8}$$

In fact $n(x_0)$ is nothing but a normal in $x_0$ at the support hyperplane of $G$ at $x_0$, this normal being directed to the exterior of $G$ and hence called an *outer normal* to bdry $G$. If for each $x_0 \in \text{bdry } G$, we define the $C^1$ function $V_{x_0}$ by

$$V_{x_0}(x) = (x - x_0) \cdot n(x_0), \quad x \in \mathbf{R}^n,$$

then $V_{x_0}(x_0) = 0$ and, by (7.8), condition (i) for a bound set holds. Moreover

$$V'_{x_0}(x) = n(x_0)$$

for each $x \in \mathbf{R}^n$ and hence we obtain from Theorem 7.4 the following

COROLLARY 7.1: *Assume that there exists an open bounded convex set* $G \subset \mathbf{R}^n$ *with* $0 \in G$ *such that, for each* $x_0 \in \operatorname{bdry} G$, *one can find an outer normal* $n(x_0)$ *to* bdry $G$ *for which*

$$n(x_0) \cdot f(t, x_0) < 0$$

*when* $t \in I$. *Then the BVP* (7.1)–(7.2) *has at least one solution* $x$ *such that* $x(t) \in \operatorname{cl} G$ *for all* $t \in I$.

To formulate another interesting special case of Theorem 7.4, we shall introduce a type of scalar function related to Lyapunov functions in stability theory. A $C^1$ function $V : \mathbf{R}^n \to \mathbf{R}$ will be called a *guiding function* for equation (7.1) if there exists $R > 0$ such that, for all $t \in I$ and all $x \in \mathbf{R}^n$ for which $|x| \geqslant R$, one has

$$V'(x) \cdot f(t, x) < 0, \tag{7.9}$$

where $V'$ is the gradient of $V$. We shall now prove

COROLLARY 7.2: *If equation* (7.1) *admits a guiding function* $V$ *such that*

$$V(x) \to \infty \quad \text{if} \quad |x| \to \infty, \tag{7.10}$$

*then the BVP* (7.1)–(7.2) *has at least one solution.*

*Proof:* Let $\rho > 0$ be such that

$$\rho > \sup_{|x| \leqslant R} V(x), \tag{7.11}$$

and let us define the set $G \subset \mathbf{R}^n$ by

$$G = \{x \in \mathbf{R}^n : V(x) < \rho\}.$$

$G$ is obviously open, $0 \in G$, and, by (7.10), $G$ is bounded. If we now define, for each $x_0 \in \operatorname{bdry} G$,

$$V_{x_0}(x) = V(x) - \rho,$$

then, for each $x_0 \in \text{bdry } G$,

$$G = \{x \in R^n : V_{x_0}(x) < 0\},$$

and $V_{x_0}(x_0) = 0$. Moreover, for each $x_0 \in \text{bdry } G$, $V(x_0) = \rho$ and hence by (7.11) $|x_0| > R$, which implies that, for all $t \in I$,

$$V'_{x_0}(x_0) \cdot f(t, x_0) = V'(x_0) \cdot f(t, x_0) < 0,$$

and hence $G$ is an attractive bound set relative to (7.1), and the result follows from Theorem 7.4.

In the special case of BVP for second order vector differential equations, specific results in the line of the qualitative method we have just described for getting *a priori* estimates have been obtained by using not only, as in Theorem 7.4, the first order necessary condition for the existence of an extremum but also the second order conditions for a maximum. We shall not describe this type of results here. Also, another technique for getting *a priori* bounds, and which could be applied also here, is the obtaining of estimates in $L^2$-norm.

Theorem 7.1 can also be used to prove existence theorems for more general BVP of the form (5.2)–(5.3) with $\varepsilon = 1$ and a non-critical linear part, but their discussion falls outside the scope of this paper.

## 8. BVP FOR STRONGLY NONLINEAR DIFFERENTIAL EQUATIONS: CRITICAL CASE

Let us still consider the equation (7.1) with the regularity assumptions of section 7 but consider now the periodic BC

$$x(0) - x(1) = 0. \tag{8.1}$$

In section 4, we have shown that the BVP (7.1)–(8.1) is equivalent to solving, in the space $X$ of continuous mappings $x : I \to \mathbf{R}^n$ verifying (8.1), an equation of the form

$$Lx = Nx, \tag{8.2}$$

where the linear mapping $L : X \to Y$ is easily shown to have the following properties:

(a) $\ker L$ has a finite dimension;

(b) $\operatorname{Im} L$ is closed and has a finite codimension;

(c) $\dim \ker L = \operatorname{codim} \operatorname{Im} L$.

A linear bounded mapping satisfying conditions (a) to (c) is called a *Fredholm mapping of index zero*, special cases of which are linear homeomorphisms and, when $X = Y$, linear compact perturbations of the identity. Also, one verifies as in section 7 that $N$ is a compact mapping. The Leray–Schauder continuation theorem has been extended as follows to equations (8.2) when $X$, $Y$ are normed spaces, $L : X \to Y$ is a Fredholm mapping of index zero and $N : \operatorname{cl} \Omega \to Y$ is a compact mapping, with $\Omega \subset X$ an open bounded set.

THEOREM 8.1: *Assume that the following conditions hold*:

(a′) $Lx \neq \lambda Nx$ *for each* $x \in \operatorname{bdry} \Omega$ *and each* $\lambda \in ]0, 1[$;

(b′) $Nx \notin \operatorname{Im} L$ *for each* $x \in \ker L \cap \operatorname{bdry} \Omega$;

(b″) $d_B[JQN|\ker L, \Omega \cap \ker L, 0] \neq 0$,

*where* $d_B[JQN|\ker L, \Omega \cap \ker L, 0]$ *denotes the Brouwer degree with respect to* $\Omega \cap \ker L$ *and* $0$ *of the restriction to* $\ker L$ *of the mapping* $JQN$, *with* $Q$: $Y \to Y$ *a bounded projector whose range is a direct summand of* $\operatorname{Im} L$ *and* $J : \operatorname{Im} Q \to \ker L$ *any isomorphism.*

*Then, for each* $\lambda \in [0, 1]$, *equation*

$$Lx = \lambda Nx$$

*has at least one solution in* $\operatorname{cl} \Omega$.

Note that the existence of $Q$ and $J$ is insured by the fact that $L$ is a Fredholm mapping of index zero.

In this theorem, which will not be proved here, assumption (a′) exactly corresponds to assumption (a) of Theorem 7.1 but assumptions (b′) and (b″) are clearly much more complicated than condition (b) of Theorem 7.1. This is nothing but the price paid for the greatest generality of Theorem 8.1 because it can be shown that, if $X = Y$ and $L = I$, conditions (b′) and (b″) reduce to (b). Condition (b″) requires the nontrivial concept of *Brouwer degree* of a continuous mapping from a set of a finite-dimensional vector space into itself which is essentially an "algebraic count" of the number of the zeroes of the mapping situated in the set. More specifically, if $g : \text{cl } A \subset E \to E$, with $E$ a finite-dimensional vector space and $A$ an open bounded set, is a $C^1$ mapping such that $0 \notin g(\text{bdry } A)$ and having all its possible zeroes in $A$ isolated (and hence in finite number $m$), the Brouwer degree of $g$ with respect to $A$ and 0 is the integer

$$d_B[g, A, 0] = \sum_{j=1}^{m} \text{sign det } g'(x_j),$$

where $g'$ is the (Frechet) derivative of $g$ and the sum is extended to the set of zeroes $x_1, \ldots, x_m$ of $g$ in $A$. In the general case of a continuous mapping such that $0 \notin g(\text{bdry } A)$, one has to use a delicate limit process to define the corresponding Brouwer degree which is still an integer.

In the case of our BVP, one easily checks using the discussion of section 4 that, if $a \in \{x \in X : x \text{ is a constant mapping}\}$ and if constant mappings are identified with their values in the space $\mathbf{R}^n$, one has

$$(QNa)(t) = t \int_0^1 f(s, a)\, ds = tF(a),$$

with $F$ defined in (4.9). Taking the isomorphism

$$J : \text{Im } Q \to \ker L, \ (\gamma : t \to tc) \mapsto c,$$

we obtain

$$JQN(a) = F(a), \tag{8.3}$$

and hence, as in the perturbational case, this mapping plays a basic role in the existence theorem for (7.1)–(8.1) deduced from Theorem 8.1. Of course, the tools introduced in section 7 to obtain *a priori* estimates for the possible solutions of (7.1)–(7.2) are still extremely useful in verifying condition (a′) of Theorem 8.1 applied to (7.1)–(8.1) and the following results can be proved quite similarly than in section 7:

THEOREM 8.2: *If there exists a bound set $G$ relative to* (7.1) *and if*

$$d_B[F, G, 0] \neq 0, \tag{8.4}$$

*then the BVP* (7.1)–(8.1) *has at least one solution $x$ such that $x(t) \in \operatorname{cl} G$, for $t \in I$.*

COROLLARY 8.1: *If there exists an open bounded convex set $G \in \mathbf{R}^n$, with $0 \in G$ such that, for each $x_0 \in \operatorname{bdry} G$ one can find an outer normal $n(x_0)$ to* bdry *$G$ for which*

$$n(x_0) \cdot f(t, x_0) \neq 0$$

*when $t \in I$, and if* (8.4) *holds, then the BVP* (7.1)–(8.1) *has at least one solution $x$ such that $x(t) \in \operatorname{cl} G$ for all $t \in I$.*

COROLLARY 8.2: *If equation* (7.1) *admits a guiding function such that* (7.10) *holds and*

$$d_B[V', B(R), 0] \neq 0,$$

*then the* BVP (7.1)–(8.1) *has at least one solution.*

Existence theorems when $f$ is subjected to growth restrictions as well as results for more general problems of the form (5.2)–(5.3) with $\varepsilon = 1$ can also be proved by the same approach but are outside of the scope of this paper.

As an example of application of Theorem 8.1, we shall prove a necessary and sufficient condition of solvability for the periodic BVP corresponding to a first order scalar equation. Thus, let us consider the problem (7.1)–(8.1) with $n = 1$ and assume that there exists continuous functions

$$f_+ : I \to \mathbf{R}, \quad f_- : I \to \mathbf{R}$$

such that

$$\lim_{x \to -\infty} f(t, x) = f_-(t), \quad \lim_{x \to \infty} f(t, x) = f_+(t) \tag{8.5}$$

uniformly in $t \in I$ and

$$f_{-}(t) < f(t, x) < f_{+}(t) \tag{8.6}$$

for all $t \in I$ and $x \in \mathbf{R}$. Assume that $x$ is a solution of (7.1)–(8.1). Then, integrating over $I$ both members of (7.1) and using (8.1), we get

$$\int_0^1 f(t, x(t))\, dt = 0$$

and hence, from (8.6),

$$\int_0^1 f_{-}(t)\, dt < 0 < \int_0^1 f_{+}(t)\, dt, \tag{8.7}$$

which is therefore a necessary condition for the solvability of (7.1)–(8.1). We shall use Theorem 8.1 to show that (8.7) is also sufficient. Let $x$ be a possible solution of

$$x' = \lambda f(t, x) \tag{8.8}$$

$$x(0) - x(1) = 0 \tag{8.9}$$

for some $\lambda \in ]0, 1[$. Then, using (8.6) and (8.8) we have

$$|x'|_X \leqslant B, \tag{8.10}$$

where

$$B = \max\left\{ \max_{t \in I} |f_{-}(t)|, \max_{t \in I} |f_{+}(t)| \right\}.$$

Let now $\eta > 0$ be such that

$$\int_0^1 f_{-}(t)\, dt \leqslant -\eta < 0 < \eta \leqslant \int_0^1 f_{+}(t)\, dt. \tag{8.11}$$

Such an $\eta$ exists by (8.7) and, using (8.5), there exists $\rho > 0$ such that, for each $t \in I$,

$$0 < f_+(t) - f(t, x) \leqslant \eta/2 \quad \text{for} \quad x \geqslant \rho,$$
$$0 < f(t, x) - f_-(t) \leqslant \eta/2 \quad \text{for} \quad x \leqslant -\rho.$$

This implies that

$$0 < \int_0^1 [f_+(t) - f(t, x)]\, dt \leqslant \eta/2 \quad \text{if} \quad x \geqslant \rho,$$

$$0 < \int_0^1 [f(t, x) - f_-(t)]\, dt \leqslant \eta/2 \quad \text{if} \quad x \leqslant -\rho,$$

and therefore, using (8.11),

$$\begin{aligned} &\int_0^1 f(t, x)\, dt \geqslant \eta/2 \quad \text{if} \quad x \geqslant \rho, \\ &\int_0^1 f(t, x)\, dt \leqslant -\eta/2 \quad \text{if} \quad x \leqslant -\rho. \end{aligned} \tag{8.12}$$

Now, for a possible solution $x$ of (8.8)–(8.9), one has

$$\int_0^1 f(t, x(t))\, dt = 0$$

and hence, by (8.12) there must necessarily exist some $\tau \in I$ for which

$$|x(\tau)| < \rho.$$

This implies, using (8.10), that, for each possible solution of (8.8)–(8.9), and for each $t \in I$,

$$|x(t)| = \left| x(\tau) + \int_\tau^t x'(s)\, ds \right| < \rho + B = R.$$

Thus condition (a′) of Theorem 8.1 holds with $\Omega = B(R)$ and by

(8.12) (b′) also holds if we note that

$$\operatorname{Im} L = \{x \subset Y : x(1) = 0\}.$$

Now, with $F$ defined above, we have, from (8.12),

$$F(a) < 0 \quad \text{if} \quad a \leqslant -\rho,$$
$$F(a) > 0 \quad \text{if} \quad a \geqslant \rho,$$

which, by elementary properties of Brouwer degree, implies that

$$d_B[F, B(R), 0] = 1$$

and achieves the proof. Thus we have the following

THEOREM 8.3: *Suppose that equation* (7.1) *with* $n = 1$ *verifies conditions* (8.5) *and* (8.6). *Then, a necessary and sufficient condition for the existence of at least one solution for* (7.1)–(8.1) *is that inequalities* (8.7) *hold.*

Results of the type of Theorem 8.3 can of course be proved for more general equations. Also, other techniques for *a priori* estimates, as, for example, the use of *estimates in* $L^2$*-norms*, can be extremely useful but will not be developed here.

## 9. HISTORICAL REMARKS AND INDICATIONS FOR FURTHER STUDIES

The *method of Poincaré* for periodic problems is developed in the books of Coddington and Levinson [7], Cronin [9], Sansone and Conti [36], Krasnosel'skii [20], Vainberg and Trenogin [40]. For a theoretical and numerical use of the *shooting method* see Bailey, Shampine, and Waltman [1], and, for various approaches in boundary value problems, see Bernfeld and Lakshmikantham [3].

For results in the line of Theorem 3.2 in the case of more general linear parts, one can consult, for periodic boundary conditions, the books of Halanay [13], Hale [14], Reissig, Sansone, and

Conti [33], Rouche and Mawhin [35] and, for other boundary conditions, Hartman [17] and Conti [8].

Various approaches, historically beginning with the work of Lyapunov [26] and Schmidt [37] on nonlinear integral equations, and usually referred as *alternative methods*, have been developed for obtaining results in the line of Theorem 4.2 (see the books of Cesari [5, 6] Hale [14, 15], Krasnosel'skii *et al.* [22], Vainberg and Trenogin [40], Rouche and Mawhin [35]).

Theorem 6.1 was first obtained, using Poincaré's method, by Loud [25] and the proof given here is new. For other boundary conditions, results in this line can be traced to Hammerstein [16] and one can consult the book of Krasnosel'skii [16].

For the Leray–Schauder fixed point theory, as well as for Brouwer degree, see the books of Berger and Berger [2], Cronin [9], Deimling [10], Schwartz [39]; a short account of the main features is given in Rouche and Mawhin [35]. Applications to BVP with growth restrictions in non-resonant cases can be found in Krasnosel'skii [21], Roseau [34], Conti [8], Reissig, Sansone, and Conti [33], Bernfeld and Lakshmikantham [3]. The concept of bound set is due to Gaines [30], as well as Theorem 7.4 for various BC. Corollary 7.1 uses techniques of Bebernes, Schmitt and Gustafson (see Schmitt [38]) and we owe guiding functions to Krasnosel'skii [20, 21] who used it together with Poincaré's method, a functional analytic treatment being given in Rouche and Mawhin [35]. See also Jackson [18] and Bernfeld and Lakshmikantham [3] for various methods in non-resonant cases for strongly nonlinear BVP, as well as Schmitt [38].

Theorem 8.1 is due to Mawhin [27] and has been applied to various situations (see, e.g., [28, 29, 30, 35] for further references). Results in the line of Corollaries 8.1 and 8.2 and of Theorem 8.2 can be found in [29, 30]. Necessary and sufficient conditions in the spirit of Theorem 8.3 for resonant cases have been initiated by Lazer and Leach [23] by a clever but technically complicated use of the Schauder fixed point theorem. Such results have received much attention (see [28, 30] for references).

"Mais je ne m'arrête point à expliquer ceci plus en détail, à cause que je vous ôterais le plaisir de l'apprendre de vous-même, et l'utilité de cultiver votre esprit en vous y exerçant, qui est à mon avis la principale de cette science".

The quotation is from Descartes, in "La Géométrie".

## REFERENCES

1 P. B. Bailey, L. F. Shampine, and P. E. Waltman, *Nonlinear Two-Point Boundary Value Problems*, Academic Press, New York, 1968.

2. M. Berger and M. Berger, *Perspectives in Nonlinearity*, Benjamin, New York, 1968.

3. S. R. Bernfeld and V. Lakshmikantham, *An Introduction to Nonlinear Boundary Value Problems*, Academic Press, New York, 1974.

4. G. D. Birkhoff and O. D. Kellogg, "Invariant points in function spaces," *Trans. Amer. Math. Soc.*, **23** (1922), 96–115.

5. L. Cesari, *Asymptotic Behavior and Stability Problems in Ordinary Differential Equations*, 3rd ed., Springer, Berlin, 1971.

6. ———, "Alternative methods in nonlinear analysis," in *Intern. Confer. Differential Equations, Los Angeles (1974)*, Academic Press, New York, 1975, 95–148.

7. E. A. Coddington and N. Levinson, *Theory of Ordinary Differential Equations*, McGraw-Hill, New York, 1955.

8. R. Conti, "Recent trends in the theory of boundary value problems for ordinary differential equations," *Boll. Un. Mat. Ital.*, (3) **22** (1967), 135–178.

9. J. Cronin, *Fixed Points and Topological Degree in Nonlinear Analysis*, Amer. Math. Soc., Providence, 1964.

10. K. Deimling, *Nichtlineare Gleichungen und Abbildungsgrade*, Springer, Berlin, 1974.

11. C. Goffman, "Preliminaries to functional analysis," *Studies in Modern Analysis*, R. C. Buck, ed., Mathematical Association of America, 1962, 138–180.

12. L. M. Graves, "Nonlinear mappings between Banach spaces," *Studies in Real and Complex Analysis*, I. I. Hirschman, Jr., ed., Mathematical Association of America, 1965, 34–54.

13. A. Halanay, *Differential Equations: Stability, Oscillations, Time Lags*, Academic Press, New York, 1966.

14. J. K. Hale, *Ordinary Differential Equations*, Wiley-Interscience, New York, 1969.

15. ———, *Applications of Alternative Problems*, Brown University Lecture Notes 71-1, Providence, 1971.

16. A. Hammerstein, "Nichtlineare Integralgleichungen nebst Anwendungen," *Acta Math.*, **54** (1930), 117–176.

17. P. Hartman, *Ordinary Differential Equations*, Wiley, New York, 1964.

18. L. K. Jackson, "Subfunctions and second-order ordinary differential inequalities," *Advances in Math.*, **2** (1968), 307–363.

19. M. A. Krasnosel'skii, *Topological Methods in the Theory of Nonlinear Integral Equations*, Pergamon Press, Oxford, 1963.

20. ———, *The Operator of Translation along the Trajectories of Differential Equations*, Amer. Math. Soc., Providence, 1968.

21. ———, "The theory of periodic solutions of non-autonomous differential equations," *Russian Math. Surveys*, **21** (1966), 53–74.

22. M. A. Krasnosel'skii, G. M. Vainikko, P. P. Zabreiko, Ya. B. Rutitskii, and V. Ya. Stetsenko, *Approximate Solutions of Operator Equations*, Wolters-Noordhoff, Groningen, 1972.

23. A. C. Lazer and D. E. Leach, "Bounded perturbations of forced harmonic oscillators at resonance," *Ann. Mat. Pura Appl.*, (4) **82** (1969), 49–68.

24. J. Leray and J. Schauder, "Topologie et équations fonctionnelles," *Ann. Sci. École Norm. Sup.*, (3) **51** (1934), 45–78.

25. W. S. Loud, "Periodic solutions of nonlinear differential equations of Duffing type," in *Proc. US-Japan Semin. on Differential and Functional Equations*, Benjamin, New York, 1967, 199–224.

26. A. M. Lyapunov, "Sur les figures d'équilibre peu différentes des ellipsoïdes d'une masse liquide homogène dotée d'un mouvement de rotation," *Zap. Akad. Nauk St. Petersbourg*, 1906, 1–225.

27. J. Mawhin, "Equivalence theorems for nonlinear operator equations and coincidence degree theory for some mappings in locally convex topological vector spaces," *J. Differential Equations*, **12** (1972), 610–636.

28. ———, "Topology and nonlinear boundary value problems," in *Dynamical Systems, vol. 1*, Academic Press, New York, 1976, 51–82.

29. ———, "Recent results on periodic solutions of differential equations," in *Intern. Confer. Differential Equations, Los Angeles (1974)*, Academic Press, New York, 1975, 537–556.

30. J. Mawhin and R. E. Gaines, *Alternative problems, coincidence degree and nonlinear differential equations*, Lecture Notes in Mathematics, vol. 568, Springer, Berlin, 1977.

31. E. Picard, "Sur l'application des méthodes d'approximations successives à l'étude de certaines équations différentielles," *J. Math.*, **9** (1893), 217–271.

32. H. Poincaré, *Les Méthodes Nouvelles de la Mécanique Céleste*, 3 volumes, Gauthier–Villars, Paris, 1892–1899.

33. R. Reissig, G. Sansone, and R. Conti, *Non-linear Differential Equations of Higher Order*, Noordhoff, Leyden, 1974.

34. M. Roseau, "Solutions périodiques ou presque-périodiques des systèmes différentiels de la mécanique non linéaire," *CISM Courses and Lect. 44, Udine*, 1970.

35. N. Rouche and J. Mawhin, *Équations Différentielles Ordinaires*, 2 volumes, Masson, Paris, 1973.

36. G. Sansone and R. Conti, *Non-linear Differential Equations*, Pergamon Press, Oxford, 1964.

37. E. Schmidt, "Zur Theorie der linearen und nichtlinearen Integralgleichungen, III," *Math. Ann.*, **65** (1908), 370–399.

38. K. Schmitt, "Randwertaufgaben für gewöhnliche Differentialgleichungen," in *Proc. Steiermark. Math. Symposium, Graz, Austria, 1973.*

39. J. T. Schwartz, *Nonlinear Functional Analysis*, Gordon and Breach, New York, 1969.

40. M. M. Vainberg and V. A. Trenogin, *Theory of Branching of Solutions of Non-linear Equations*, Noordhoff, Leyden, 1974.

# FIXED POINT THEOREMS AND ORDINARY DIFFERENTIAL EQUATIONS

*H. A. Antosiewicz*

Fixed point theorems play an important role in the theory and the applications of ordinary differential equations.

In this article we discuss a few simple examples of the use of the classical theorems of Banach, Brouwer, and Schauder. Our aim is to present the underlying basic ideas from a unified point of view. The setting we have chosen is essentially self-contained and elementary yet still general enough, we hope, to offer a glimpse of the broad and exciting vistas that lie beyond.

No attempt at completeness has been made in citing the references listed at the end.

## I. BANACH'S FIXED POINT THEOREM AND SOME OF ITS APPLICATIONS

**1.** There are few results as basic, and as simple to state as Banach's contraction mapping principle [7].

This work was done with partial support from the U.S. Army Research Office (Durham) under grant DAHC04-74-G-0013.

(1.1) *Let $E$ be a Banach space, let $U \subset E$ be an open ball with center $z$ and radius $r$, and suppose $T$: $U \to E$ is a strict contraction, i.e., there is a constant $k \in [0, 1[$ such that*

$$|Tu_1 - Tu_2| \leqslant k|u_1 - u_2| \tag{1.1.1}$$

*for any $u_1$, $u_2$ in $U$. If*

$$|Tz - z| < (1 - k)r, \tag{1.1.2}$$

*there exists a unique point $v \in U$ such that $Tv = v$.*

Indeed, the assumptions imply that $T(U) \subset U$, and that $(T^n z)$ is a Cauchy sequence in the closed ball with center $z$ and radius $|Tz - z|(1 - k)^{-1}$. Hence the latter contains a point $v$ such that

$$v = \lim T^n z = \lim T(T^n z) = Tv. \tag{1.2}$$

Clearly, $v \in U$, and $v$ is unique because of (1.1.1).

The applications of (1.1) to problems involving a differential equation of the form

$$\dot{x} = f(t, x), \tag{1.3}$$

however diverse, have one feature in common: they depend on replacing the original problem by an analogous problem for an integral equation so chosen that its solutions are equivalent to the desired solutions of (1.3). This often can be done simply and in a very natural way.

**2.** An important example is Cauchy's celebrated existence and uniqueness theorem (cf., for example, [32]).

(2.1) *Let $I \subset \mathbf{R}$ be an interval, let $H \subset \mathbf{R}^n$ be an open ball with center $z$ and radius $r$, and suppose $f$: $I \times H \to R^n$ is a continuous lipschitzian mapping, i.e., there is a constant $\lambda \in [0, 1[$ such that*

$$|f(t, x_1) - f(t, x_2)| \leqslant \lambda|x_1 - x_2| \tag{2.1.1}$$

*for any $t \in I$ and any $x_1$, $x_2$ in $H$.*

*Given any $a \in I$ and any compact set $A \subset H$, there is a compact interval $J_a \subset I$ containing $a$ such that, for every $y \in A$, there exists a unique solution $v$ of* (1.3) *in $J_a$ for which $v(a) = y$.*

A solution $v$ of (1.3) in an interval $J_a$ containing $a$, for which $v(a) = y$, is a continuously differentiable mapping of $J_a$ into $H$ such that

$$\dot{v}(t) = f(t, v(t)) \tag{2.2}$$

for every $t \in J_a$ and hence, by integration,

$$v(t) = y + \int_a^t f(s, v(s))\, ds \tag{2.3}$$

for every $t \in J_a$. In turn, any continuous mapping $v\colon J_a \to H$ that satisfies (2.3) at each $t \in J_a$ is in fact continuously differentiable and a solution of (1.3) in $J_a$ for which $v(a) = y$. Thus it suffices to prove the existence of a solution of (2.3).

Let $J \subset I$ be a compact interval containing $a$, choose $M > 0$ so that $|f(t, z)| \leqslant M$ for every $t \in J$, and let $J_a \subset J$ be a compact interval, containing $a$ and of length

$$\Delta < (\lambda r + M)^{-1} d(A, \complement H) \tag{2.4}$$

so that in particular, $\lambda\Delta < 1$.

Let $E$ be the Banach space of continuous mappings of $J_a$ into $\mathbf{R}^n$, and denote by $U \subset E$ the open ball with center $z$ and radius $r$.* Clearly, $f$ which maps $J_a \times H$ into $\mathbf{R}^n$ induces a mapping $F$ of $U$ into $E$ by setting, for each $u \in U$,

$$Fu(t) = f(t, u(t)) \tag{2.5}$$

at every $t \in J_a$; $F$ is often called the Niemitzky operator generated by $f$.

*As usual, we identify constant mappings in $E$, such as $t \mapsto z$, to their values in $\mathbf{R}^n$.

Similarly, integration of any $u \in E$ may be viewed as a mapping $K$ of $E$ into itself by setting, for each $u \in E$,

$$Ku(t) = \int_a^t u(s)\, ds \tag{2.6}$$

at every $t \in J_a$.

Thus, for any $y \in A$,

$$T = y + KF \tag{2.7}$$

is a mapping of $U$ into $E$, and any $v \in U$ satisfies (2.3) if and only if $v = Tv$. As a result, the assertions in (2.1) will be a direct consequence of (1.1) provided $T$ is a strict contraction satisfying (1.1.2).

It is immediate that (1.1.1) holds with $k = \lambda\Delta$ where $k \in [0, 1[$ because of (2.4). In fact, (1.1.2) also holds because (2.4) implies, for any $y \in A$,

$$|Tz - z| \leqslant |y - z| - kr + \Delta(\lambda r + M) < (1 - k)r. \tag{2.8}$$

Observe that the assertions in (2.1) can be strengthened somewhat: If $v_1$, $v_2$ are two solutions of (1.3) in $J_a$ for which $v_i(a) = y_i$, $i = 1, 2$, then

$$|v_1 - v_2| = |y_1 - y_2 + KFv_1 - KFv_2| \leqslant (1 - k)^{-1}|y_1 - y_2| \tag{2.9}$$

because of (2.7). This shows the (uniform) continuity in $A$ of the mapping of $A$ into $E$ that associates to each $y \in A$ the solution of (1.3) in $J_a$ which equals $y$ at $a$.

**3.** Similar considerations can be applied in more general, and quite different circumstances.

Consider a differential equation of the form

$$\dot{x} = A(t)x + f(t, x) \tag{3.1}$$

where $A(t)$, for each $t \in \mathbf{R}_+$, is an endomorphism of $\mathbf{R}^n$ such that

$t \mapsto A(t)$ is continuous in $\mathbf{R}_+$, and $f$ is a continuous lipschitzian mapping of a set $\mathbf{R}_+ \times H_0$ into $\mathbf{R}^n$, $H_0 \subset \mathbf{R}^n$ being an open ball of radius $r_0$ centered at the origin.

It is easy to verify that a continuous mapping $v$ of an interval $J = [a, b[ \subset \mathbf{R}_+$ into $H_0$ is a solution of (3.1) in $J$ that equals $y$ at $a$ if and only if $v$ is a solution in $J$ of the (non-homogeneous) linear differential equation

$$\dot{x} = A(t)x + f(t, v(t)) \tag{3.2}$$

or equivalently, if and only if for every $t \in J$,

$$v(t) = R(t)R^{-1}(a)y + \int_a^t R(t)R^{-1}(s)f(s, v(s))\, ds \tag{3.3}$$

where $R$ is the principal fundamental (matrix) solution in $\mathbf{R}_+$ of the linear differential equation

$$\dot{x} = A(t)x. \tag{3.4}$$

As is well known, $R(t)$ is an invertible endomorphism of $\mathbf{R}^n$ for each $t \in \mathbf{R}_+$ and hence, for any $(a, y) \in \mathbf{R}_+ \times \mathbf{R}^n$, the (unique) solution of (3.4) in $\mathbf{R}_+$ that equals $y$ at $a$ is given by $t \mapsto R(t)R^{-1}(a)y$. Indeed, (3.3) represents the classical "variation of constants" formula for the explicit solution of (3.2).

Of course, the nonlinear differential equation (3.1) need not admit a solution in $[a, +\infty[$ for any $a \in \mathbf{R}_+$. Sufficient conditions in order that it does may be stated as follows [21], [44]:

(3.5) *Let $A(t)$, for each $t \in \mathbf{R}_+$, be an endomorphism of $\mathbf{R}^n$ such that $t \mapsto A(t)$ is continuous in $\mathbf{R}_+$, and suppose there is a projection $P$ in $\mathbf{R}^n$ and a constant $\Delta > 0$ such that*

$$\int_0^t |R(t)PR^{-1}(s)|\, ds + \int_t^\infty |R(t)[I - P]R^{-1}(s)|\, ds \leqslant \Delta \tag{3.5.1}$$

*for every $t \in \mathbf{R}_+$.*

*Let $f\colon \mathbf{R}_+ \times H_0 \to \mathbf{R}^n$ be a continous mapping which is lipschitzian for a constant $\lambda \in [0, \Delta^{-1}[$ and satisfies $f(t, 0) = 0$ for* *every* $t \in \mathbf{R}_+$, $H_0 \subset \mathbf{R}^n$ *being an open ball of radius* $r_0$ *centered at the origin. Then there exists a constant* $\alpha > 0$ *such that, for every* $y \in \operatorname{im} P$ *with* $|y| < (1 - \Delta\lambda)r_0/\alpha$, *there exists a unique solution* $v$ *of* (3.1) *in* $\mathbf{R}_+$ *for which* $Pv(0) = y$ *and*

$$|v(t)| < r_0 \tag{3.5.2}$$

*for every* $t \in \mathbf{R}_+$.

Observe that the assumptions (3.5.1) are verified, for example, whenever $A(t)$ is independent of $t$ in $\mathbf{R}_+$ and has no purely imaginary eigenvalue. Actually, in that case, a sharper result can be proved [21], [45].

For the proof of (3.5) note first that (3.5.1) implies, in particular, that each of the integrals does not exceed $\Delta$ at any $t \in \mathbf{R}_+$. This in turn implies, on the one hand, the existence of a constant $\alpha > 0$ such that, given any $y \in \operatorname{im} P$,

$$|R(t)y| \leqslant \alpha|y|\exp(-t/\Delta) \tag{3.6}$$

for every $t \in \mathbf{R}_+$ and, on the other hand, that a solution $v_0$ of (3.4) in $\mathbf{R}_+$ is bounded if and only if $v_0(0) = Pv_0(0)$ (in which case (3.6) holds with $y = v_0(0)$).

Thus, a continuous mapping $v$ of $\mathbf{R}_+$ into $\mathbf{R}^n$ satisfying (3.5.2) will be a (bounded) solution of (3.1) in $\mathbf{R}_+$ for which $Pv(0) = y$ if and only if $v$ is a bounded solution of (3.2) in $\mathbf{R}_+$ with $Pv(0) = y$ or, equivalently, because of (3.3) and the hypotheses in (3.5), if and only if

$$\begin{aligned} v(t) = R(t)y &+ \int_0^t R(t)PR^{-1}(s)f(s, v(s))\,ds \\ &- \int_t^\infty R(t)[I - P]R^{-1}(s)f(s, v(s))\,ds \end{aligned} \tag{3.7}$$

for every $t \in \mathbf{R}_+$ and, in particular,

$$(I - P)v(0) + \int_0^\infty (I - P)R^{-1}(s)f(s, v(s))\,ds = 0. \tag{3.8}$$

Note that $v_0: t \mapsto R(t)y$ is the solution of (3.4) in $\mathbf{R}_+$ for which $v_0(0) = Pv_0(0)$ $(= y)$ and hence $|v_0| \leqslant \alpha|y|$ by (3.6).

The equation (3.7) suggests the following setting for the application of (1.1):

Let $E$ be the Banach space of bounded continuous mappings of $\mathbf{R}_+$ into $\mathbf{R}^n$, with the norm of any $u \in E$ defined by $|u| = \sup\{|u(t)| : t \in \mathbf{R}_+\}$, as usual; let $U_0 \subset E$ be the open ball of radius $r_0$ centered at the origin; introduce as in (2.5) the Niemitzky operator $F$ generated by $f$; and define a mapping $K$ of $E$ into itself by setting, for each $u \in E$,

$$Ku(t) = \int_0^t R(t)PR^{-1}(s)u(s)\,ds$$
$$-\int_t^\infty R(t)[I - P]R^{-1}(s)u(s)\,ds \tag{3.9}$$

at every $t \in \mathbf{R}_+$. By virtue of the assumptions in (3.5), $F$ maps $U_0$ into $E$, is lipschitzian for the constant $\lambda$, and $F0 = 0$, and $K$ is linear and $|Ku| \leqslant \Delta|u|$ for every $u \in E$.

Given $y \in \operatorname{im} P$, let $v_0$ be defined as above, so that $v_0 \in E$, and put

$$T = v_0 + KF. \tag{3.10}$$

Then $T$ is a mapping of $U_0$ into $E$ which is lipschitzian for the constant $k = \Delta\lambda < 1$ and such that $|T0| \leqslant |v_0| \leqslant \alpha|y| < (1 - k)r_0$ whenever $y \in \operatorname{im} P$ satisfies $|y| < \alpha^{-1}(1 - k)r_0$. Clearly, by construction, any $v \in U_0$ for which $Tv = v$ will satisfy (3.7) at every $t \in \mathbf{R}_+$. Therefore, the assertions in (3.5) are a direct consequence of (1.1). The result (3.5) is but one of many similar results for the nonlinear differential equation (3.1) that depend on what have been called a dichotomy and an exponential dichotomy of the solutions of the linear differential equation (3.4). These notions play an important role in stability theory, where they originated (see, for example, [21], [32], [45]).

**4.** A broad class of problems for a nonlinear differential equation of the form

$$\dot{x} = A(t)x + f(t, x) \tag{4.1}$$

is concerned with the existence of solutions that satisfy additional preassigned conditions. The result (3.5) is one illustration. Another important special case is the classical problem of the existence of periodic solutions.

Let us recall a few simple facts concerning a linear differential equation and its adjoint,

$$\dot{x} = A(t)x, \quad \dot{y} = -A^*(t)y \tag{4.2}$$

when $A(t)$, for each $t \in R$, is an endomorphism of $\mathbf{R}^n$ such that $t \mapsto A(t)$ is continuous and of period $p(> 0)$. Neither equation need admit a nontrivial solution of period $p$, of course; but if one does, so does the other. In fact, in that case, the linear spaces of their solutions with period $p$ are isomorphic and each is a topological supplement of the Banach space $E$ of continuous mappings of $\mathbf{R}$ into $\mathbf{R}^n$ that are of period $p$; i.e., there are continuous projections $P$, $P_*$, respectively, such that

$$E = \ker P + \operatorname{im} P, \quad E = \ker P_* + \operatorname{im} P_*. \tag{4.3}$$

Moreover, the linear differential equation

$$\dot{x} = A(t)x + b(t), \tag{4.4}$$

for any $b \in E$, has a solution of period $p$ if and only if $b \in \ker P_*$, and if $b \in \ker P_*$, there is one and only one solution of (4.4) that belongs to $\ker P$. Indeed, if $K$ denotes that mapping of $\ker P_*$ into $\ker P$, then $K$ is linear and $K(I_E - P_*)$ is a mapping of $E$ into $\ker P$ which is linear and continuous.†

Analogous arguments apply to the nonlinear differential equation (4.1). More precisely, suppose $f$ in (4.1) is a continuous mapping of $\mathbf{R} \times H_0$ into $\mathbf{R}^n$ such that $f(t + p, x) = f(t, x)$ for

---

†Here, and in the sequel, $I_E$ and $I$ will denote the identity mappings in $E$ and $\mathbf{R}^n$, respectively.

every $(t, x) \in \mathbf{R} \times H_0$, $H_0 \subset \mathbf{R}^n$ being an open ball of radius $r_0$ centered at the origin; and let $F$ again be the Niemitzky operator induced by $f$, as in (2.5), so that $F$ is a continuous mapping into $E$, of the open ball $U_0 \subset E$ with center at the origin and radius $r_0$.

Clearly, a solution $v$ of (4.1) in $\mathbf{R}$ with period $p$ is a solution of period $p$ of (4.4) with $b = Fv$; therefore, necessarily, $Fv \in \ker P_*$ and $v - Pv = KFv$. Thus, for every isomorphism $L$ of im $P_*$ onto im $P$

$$v = Pv + LP_*Fv + K(I_E - P_*)Fv. \tag{4.5}$$

Conversely, given any isomorphism $L$ of im $P_*$ onto im $P$

$$S_L = P + LP_*F + K(I_E - P_*)F \tag{4.6}$$

is a continuous mapping of $U_0$ into $E$ such that every $v \in U_0$ for which $v = S_L v$ is a solution of (4.1) in $\mathbf{R}$ with period $p$.

Two special cases of (4.1) or, equivalently, of (4.6) are of particular interest: when (4.2) admits no nontrivial solution of period $p$, and when $A(t) = 0$ for every $t \in \mathbf{R}$ (cf., for example, [2], [3], [27], [32], [50]).

(4.7) *Let $A(t)$, for every $t \in \mathbf{R}$, be an endomorphism of $\mathbf{R}^n$ such that $t \mapsto A(t)$ is continuous and of period $p(> 0)$, and suppose that*

$$\dot{x} = A(t)x \tag{4.7.1}$$

*has no nontrivial solution of period $p$.*

*Let $f: \mathbf{R} \times H_0 \to \mathbf{R}^n$ be a continuous mapping which is lipschitzian for a constant $\lambda > 0$ and satisfies $f(t + p, x) = f(t, x)$ for every $(t, x) \in \mathbf{R} \times H_0$, $H_0 \subset \mathbf{R}^n$ being an open ball of radius $r_0$ centered at the origin.*

*Then there exists a constant $\Delta > 0$ such that, if $k = \Delta\lambda < 1$ and $|f(t, 0)| \leqslant \Delta^{-1}(1 - \Delta\lambda)r_0$ for every $t \in \mathbf{R}$, there exists a unique solution $v$ of (4.1) in $\mathbf{R}$ which is of period $p$.*

The assumptions imply that $P = P_* = 0$ and that $I - R(p)$ is invertible, where $R$ is the principal fundamental (matrix) solution of (4.7.1). Thus $K$ is the mapping of $E$ into itself which is defined, for each $u \in E$, by

$$Ku(t) = \int_0^p G(t, s)u(s)\, ds \tag{4.8}$$

at every $t \in \mathbf{R}$, where $G(t, s)$ denotes the so-called Green's operator

$$G(t, s) = \begin{cases} R(t)\left[I - R(p)\right]^{-1} R^{-1}(s), & 0 \leqslant s \leqslant t \leqslant p, \\ R(t + p)\left[I - R(p)\right]^{-1} R^{-1}(s), & 0 \leqslant t < s \leqslant p. \end{cases} \tag{4.9}$$

Clearly, there is a $\Delta > 0$ such that $|Ku| \leqslant \Delta|u|$ for every $u \in E$.

It follows that (4.6) reduces to $S_0 = KF$, and that $S_0$ is a strict contraction of $U_0$ into $E$ provided $k = \Delta\lambda < 1$. Hence the assertion in (4.7) is a direct consequence of the assumptions and of (1.1).

When $A(t) = 0$ for each $t \in \mathbf{R}$, so that (4.1) becomes

$$\dot{x} = f(t, x), \tag{4.10}$$

it is easy to see that both $P$ and $P_*$ coincide with the mapping $P_0$ that associates with each $u \in E$ the constant mapping equal to the mean value of $u$, i.e.,

$$P_0 u \colon t \mapsto \frac{1}{p} \int_0^p u(s)\, ds. \tag{4.11}$$

Moreover, the mapping $K$, which by definition maps $\ker P_0$ into itself, reduces to the mapping $K_0$ that associates to each $u \in E$ with $P_0 u = 0$ the primitive of $u$ with mean value 0; i.e., for each $u \in \ker P_0$,

$$K_0 u(t) = \int_0^t u(s)\, ds - \frac{1}{p} \int_0^p \int_0^t u(s)\, ds\, dt \tag{4.12}$$

at every $t \in \mathbf{R}$.

Thus, in this case, (4.6) can be written in the form

$$S_L = P_0 + LP_0F + K_0(I - P_0)F \tag{4.13}$$

where $L$ is simply an isomorphism in $\mathbf{R}^n$, and the question of the existence of a solution of (4.10) with period $p$ is equivalent to the problem of the existence of a fixed point of the mapping (4.13).

The same considerations that lead to (4.13), in this particular case, yield yet another equivalent formulation. Indeed, the fact that any $v \in U_0$ is a solution of (4.10) with period $p$ if and only if

$$\begin{aligned} v &= P_0v + K_0Fv \\ 0 &= P_0Fv \end{aligned} \tag{4.14}$$

suggests the introduction of a mapping $\Sigma$ of $U_0 \times \mathbf{R}^n$ into $E \times \mathbf{R}^n$ such that, for every $(u, x) \in U_0 \times \mathbf{R}^n$,

$$\Sigma(u, x) = (x + K_0(I - P_0)Fu, x + P_0Fu). \tag{4.15}$$

For then, every $(v, y) \in U_0 \times \mathbf{R}^n$ for which $\Sigma(v, y) = (v, y)$ clearly satisfies (4.14) with $P_0v = y$ (and conversely). Moreover, the continuity of $\Sigma$ and such properties as being a strict contraction are again direct consequences of corresponding properties of $f$.

A wide variety of techniques has been used to attack these problems. A general approach for the solution of (4.13) based on projection methods is given in [8], [18]; for another using degree theory, see [22], [50] (cf., also, [4], [5], [46]). A special case of (4.15) is discussed below in (II. 3.10).

## II. SCHAUDER'S FIXED POINT THEOREM AND SOME OF ITS APPLICATIONS

**1.** A fixed point theorem far more sophisticated than Banach's contraction mapping principle is the following theorem of Schauder [53], which was generalized to linear, locally convex topological spaces by Tychonov [54]:

(1.1) *Let $E$ be a Banach space, let $C \subset E$ be a closed convex set, and suppose $T$ is a continuous mapping of $C$ into itself such that $T(C)$ has compact closure. Then there is a point $v \in C$ such that $Tv = v$.*

The proof requires Brouwer's fixed point theorem (cf., for example, [24]).

The applications of (1.1) to problems involving a differential equation of the form

$$\dot{x} = f((t, x), \tag{1.2}$$

like those of (I. 1.1), depend essentially upon the conversion of the original problem into a similar problem for a corresponding integral equation.

**2.** Peano's classical existence theorem provides a simple illustration of the use of (1.1) (cf., for example, [32]).

(2.1) *Let $J \subset \mathbf{R}$ be a compact interval, let $H \subset \mathbf{R}^n$ be an open set, and suppose $f$ is a continuous mapping of $J \times H$ into $\mathbf{R}^n$.*

*Given any $a \in J$ and any compact set $A \subset H$, there is a compact interval $J_a \subset J$ containing $a$ such that, for every $y \in A$, there exists at least one solution $v$ of* (1.2) *in $J_a$ for which $v(a) = y$.*

Given a compact set $A \subset H$, the continuity of $f$ in $J \times H$ implies that there is an open neighborhood $V \subset H$ of $A$ and a constant $M > 0$ such that $|f(t, x)| \leqslant M$ for all $(t, x) \in J \times V$. Let $r > 0$, $\Delta > 0$ be constants such that

$$M\Delta \leqslant r < d(A, \complement V). \tag{2.2}$$

Given any $a \in J$ and any $y \in A$, let $J_a \subset J$ be a compact interval of length $\Delta$, containing $a$; denote by $E$ the Banach space of continuous mappings of $J_a$ into $\mathbf{R}^n$; and let $U \subset E$ be the open ball with center $y$ and radius $r$, so that $u \in \overline{U}$ implies $|u(t) - y| \leqslant r$ and hence $u(t) \in V$ for each $t \in J_a$.

As before, let $F$ be the Niemitzky operator (I. 2.5) induced by $f$, which maps $\bar{U}$ into $E$; denote by $K: E \to E$ the mapping (I. 2.6), which associates with each $u \in E$ the primitive of $u$ that equals 0 at $a$; and define

$$T = y + KF, \tag{2.3}$$

which formally agrees with (I. 2.7).

It follows at once that $T$ is a continuous mapping of $\bar{U}$ into $E$, and that $v \in \bar{U}$ is a solution of (1.2) in $J_a$ with $v(a) = y$ if and only if $Tv = v$. Thus, the assertion in (2.1) will result directly from (1.1) if $T$ maps $\bar{U}$ into itself and $T(\bar{U})$ is relatively compact.

By construction, for each $u \in \bar{U}$,

$$|Tu - y| \leqslant \Delta|Fu| \leqslant \Delta M \leqslant r$$

which establishes, on the one hand, that $T(\bar{U}) \subset \bar{U}$ and, on the other, that the set $\{Tu(t): u \in \bar{U}\}$, for each $t \in J_a$, is relatively compact in $\mathbf{R}^n$. Moreover, for any $u \in \bar{U}$,

$$|Tu(t) - Tu(s)| \leqslant \left|\int_s^t |f(\tau, u(\tau))|\, d\tau\right| \leqslant M|t - s|$$

whatever $t, s$ in $J_a$, which shows that $T(\bar{U})$ is an equicontinuous set in $E$. Thus, by Ascoli's theorem, $T(\bar{U})$ is relatively compact.

Note that the finite dimensionality of $\mathbf{R}^n$ is crucial to showing that $T(\bar{U})$ is relatively compact.

The continuity of $f$ alone does not guarantee the existence of a solution of (1.2) when the underlying coordinate space is a non-reflexive Banach space of infinite dimension or, for example, the (reflexive) Hilbert space $l_2$ [17], [55] (see also, (IV. 2.4)).

**3.** Schauder's theorem (1.1) has found many diverse applications in problems dealing with the existence of periodic solutions of a differential equation of the form

$$\dot{x} = A(t)x + f(t, x). \tag{3.1}$$

This is particularly true when assumptions are made such as in

(I. 4.7) which stipulate directly, or otherwise imply that the corresponding linear differential equation has no nontrivial periodic solution.

A special case of importance is the second order differential equation

$$\ddot{x} = f(t, x) \tag{3.2}$$

and the problem of the existence of a solution $v$ of (3.2) in a given interval $J = [0, p]$ that satisfies the boundary conditions

$$v(0) = 0, \quad v(p) = 0. \tag{3.3}$$

Here, for simplicity's sake, $f$ is assumed to be a continuous mapping of $J \times H_0$ into $\mathbf{R}^n$, where $H_0 \subset \mathbf{R}^n$ is an open ball of radius $r_0$ centered at the origin.

The function space setting is established at once. For it is easy to verify that a solution $v$ of (3.2) in $J$ for which (3.3) holds satisfies, at every $t \in J$,

$$v(t) = -\int_0^p G(t, s) f(s, v(s))\, ds \tag{3.4}$$

where $G: J \times J \to \mathbf{R}$ is the Green's function defined by

$$G(t, s) = \begin{cases} \dfrac{1}{p}(p - t)s, & 0 \leqslant s \leqslant t \leqslant p, \\ \dfrac{1}{p} t(p - s), & 0 \leqslant t \leqslant s \leqslant p, \end{cases} \tag{3.5}$$

and, conversely, any continuous mapping $v: J \to \mathbf{R}^n$ with $|v(t)| < r_0$ at every $t \in J$, for which (3.4) holds, is in fact twice continuously differentiable and a solution of (3.2) in $J$ satisfying (3.3).

Thus, let $E$ be the Banach space of continuous mappings of $J$ into $\mathbf{R}^n$, let $U_0 \subset E$ be the open ball of radius $r_0$ centered at the origin, and introduce as in (I. 2.5) the Niemitzky operator $F$ generated by $f$, which maps $U_0$ into $E$, and the mapping $K: E \to E$

defined, for each $u \in E$, by

$$Ku(t) = -\int_0^p G(t, s)u(s)\,ds \tag{3.6}$$

at every $t \in J$. Then $F$ is continuous in $U_0$, $K$ satisfies $|Ku| \leqslant p^2|u|/8$ for each $u \in E$, and

$$T = KF \tag{3.7}$$

is a continuous mapping of $U_0$ into $E$ such that every fixed point $v \in U_0$ of $T$ is a solution of (3.2) satisfying (3.3).

This yields immediately the following simple result [27], [32]:

(3.8) *Let $J = [0, p] \subset \mathbf{R}$, let $H_0 \subset \mathbf{R}^n$ be an open ball of radius $r_0$ centered at the origin, and suppose $f$: $J \times H_0 \to \mathbf{R}^n$ is a continuous mapping for which there exist a closed ball $\bar{V} \subset H_0$ of radius $r$ centered at the origin and a positive constant $M \leqslant 8rp^{-2}$ such that $|f(t, x)| \leqslant M$ at every $(t, x) \in J \times \bar{V}$.*

*Then there exists at least one solution $v$ of* (3.2) *in $J$ which satisfies the boundary conditions* (3.3).

Indeed, if $C \subset E$ is the closed ball of radius $r$ centered at the origin, the assumptions imply that $T(C) \subset C$ and that, for any $u \in C$,

$$|Tu(t) - Tu(s)| \leqslant \frac{p}{2} M|t - s| \tag{3.9}$$

whatever $t$, $s$ in $J$, as a consequence of the definition (3.6) and the mean value theorem. Thus, Ascoli's theorem shows that $T(C)$ is relatively compact and hence the assertion follows directly from (1.1).

Another result in the same spirit but based on the consideration leading to (I. 4.15) may be stated in the following way [19]:

(3.10) *Let $f$ be a continuous real valued function in $\mathbf{R}^2$ which has the following properties*: (i) *there is a constant $p > 0$ such that $f(t + p, x) = f(t, x)$ for all $(t, x) \in \mathbf{R}^2$*; (ii) *there is a constant*

$r > 0$ *such that* $xf(t, x) \geqslant 0$ *for all* $(t, x) \in \mathbf{R}^2$ *with* $|x| \geqslant r$; *and* (iii)

$$\lim_{|x|\to\infty} f(t, x)/x = 0 \tag{3.10.1}$$

*uniformly in* $\mathbf{R}$.

*Then there exists at least one solution of* (3.2) *in* $\mathbf{R}$ *which has period* $p$.

The setting here involves the Banach space of continuous real valued functions in $\mathbf{R}$ with period $p$, which we denote by $E$; the mappings $K$ and $F$ corresponding to (2.6) and the Niemitzky operator induced by $f$, respectively, which are continuous mappings of $E$ into itself; and as in (I. 4.11), the projection $P_0$ in $E$ which associates to each $u \in E$ the constant mapping equal to the mean value of $u$.

As the considerations preceding (I. 4.15) show, any fixed point of the mapping $\Sigma$ of $E \times \mathbf{R}$ into itself, defined by

$$\Sigma(u, x) = (x + K(I - P_0)Fu, x + P_0Fu) \tag{3.11}$$

at every $(u, x) \in E \times \mathbf{R}$, gives rise to a solution of (3.2) in $\mathbf{R}$ which satisfies the boundary conditions (3.3) and hence has period $p$. Thus, the assertion in (3.10) will be a direct consequence of (1.1) provided there is a closed convex set $C \subset E \times \mathbf{R}$ relative to which $\Sigma$ verifies the assumptions in (1.1). Such a set is readily constructed.

Given a positive constant $k < \frac{1}{6}\min(2, p^{-2})$, there are positive constants $m$, $M$ such that $|f(t, x)| \leqslant K|x|$ for all $(t, x) \in \mathbf{R}^2$ with $|x| \geqslant m$ and $|f(t, x)| \leqslant M$ for all $(t, x) \in \mathbf{R}^2$ with $|x| \leqslant m$. Put

$$r_0 = \max\left\{\frac{r}{1 - 3k}, \frac{r}{1 - 6kp^2}, \frac{M}{k}, m\right\}, \quad p = kr_0 \max(1, 2p^2), \tag{3.12}$$

let $\overline{U}_0 \subset E$ be the closed ball of radius $r_0$ centered at the origin, let $\bar{H}_0 \subset R$ be the closed ball of radius $r + 2p$, and define

$$C = \bar{U}_0 \times \bar{H}_0.$$

Then $C$ is a closed convex subset of $E \times \mathbf{R}$, and $\Sigma(C) \subset C$ as is easily verified. Moreover, the properties of $P_0$ and $K$ imply, by Ascoli's theorem, that $\Sigma(C)$ is relatively compact. Therefore, by (1.1), $\Sigma$ has a fixed point $(v, y) \in C$, and $v$ is the desired solution of (3.2) in $\mathbf{R}$ with period $p$. Observe that, by construction,

$$y = \frac{1}{p}\int_0^p v(s)\,ds, \quad 0 = \frac{1}{p}\int_0^p f(s, v(s))\,ds. \tag{3.13}$$

**4.** The product space setting provides a particularly effective framework for dealing with many different boundary value problems for compact intervals in a unified way (cf., for example, [5]).

As an illustration, consider the differential equation

$$\dot{x} = f(t, x) \tag{4.1}$$

where $f: J \times \mathbf{R}^n \to \mathbf{R}^n$ is a continuous mapping and $J$, as before, denotes the compact interval $[0, p] \subset \mathbf{R}_+$, and suppose a solution $v$ of (4.1) in $J$ is desired which should satisfy a set of conditions of the form

$$\Gamma v = z \tag{4.2}$$

where $\Gamma$ is a continuous linear mapping into $\mathbf{R}^n$, of the Banach space $E$ of continuous mappings of $J$ into $\mathbf{R}^n$, and $z \in \mathbf{R}^n$ is a given point. $\Gamma$ may be given, for example, in terms of two endomorphisms $M$, $N$ of $\mathbf{R}^n$ by prescribing $\Gamma u = Mu(0) + Nu(p)$ for each $u \in E$. Thus, evidently, these problems include, as special case, the problem of the existence of a periodic solution.

It is clear that there exists a solution $v$ of (4.1) in $J$ satisfying (4.2) if and only if

$$(v, x) - (x + KFv, x - \Gamma v) = (0, z) \tag{4.3}$$

where, as before, $F$ is the Niemitzky operator (I. 2.5) induced by $f$ in (4.1), and $K$ is the mapping (I. 2.6) which associates to each $u \in E$ the primitive of $u$ that equals 0 at 0. This suggests at once introducing the mapping

$$\Psi: (u, x) \rightarrow (u - x - KFu, \Gamma u) \tag{4.4}$$

of $E \times \mathbf{R}^n$ into itself and inquiring into conditions under which the equation

$$\Psi(u, x) = (0, z) \tag{4.5}$$

admits a solution, for a given $z \in \mathbf{R}^n$.

If, in particular, $\Psi$ is a continuous injection such that $I - \Psi$ maps bounded sets in $E \times \mathbf{R}^n$ into relatively compact sets, then Schauder's theorem on the invariance of domain implies that $\Psi$ is a homeomorphism of $E \times \mathbf{R}^n$ onto $\Psi(E \times \mathbf{R}^n)$. Hence, if $\Psi(E \times \mathbf{R}^n)$ in that case can be shown to be closed, then necessarily $\Psi(E \times \mathbf{R}^n) = E \times \mathbf{R}^n$ and, consequently, the equation (4.5) will have a unique solution for every $z \in \mathbf{R}^n$.

Various conditions are known under which a homeomorphism of a Banach space into itself is surjective. The following is of particular interest here [5]:

(4.6) *Let $H$ be an upper semi-continuous mapping of a Banach space $X$ into the family of non-empty closed convex subsets of $X$ and suppose $h_0$. $X \rightarrow X$ is a continuous mapping that maps bounded sets into relatively compact sets. If, for every $x, y$ in $X$,*

$$h_0(x) - h_0(y) \in H(x - y) \tag{4.6.1}$$

*and if $h = I_x - h_0$ is injective, then $h$ is a homeomorphism of $X$ onto itself.*

To see the significance of the assumptions in (4.6) suppose that $f: J \times \mathbf{R}^n \rightarrow \mathbf{R}^n$ in (4.1) satisfies a condition such as

$$|f(t, x) - f(t, y)| \leqslant \lambda(t)|x - y| \tag{4.7}$$

for every $t \in J$ and every $x, y$ in $\mathbf{R}^n$, where $\lambda$ is a continuous positive valued function in $J$. Then we can define, for every $(t, x) \in J \times \mathbf{R}^n$, the set

$$L(t, x) = \{y \in \mathbf{R}^n : |y| \leqslant \lambda(t)|x|\} \tag{4.8}$$

so that (4.7) can be written in the equivalent form

$$f(t, x) - f(t, y) \in L(t, x - y). \tag{4.9}$$

Obviously, $L$ is a mapping of $J \times \mathbf{R}^n$ into the family of closed convex subsets of $\mathbf{R}^n$. Moreover, $L$ induces a mapping $H$ of $E \times \mathbf{R}^n$ into the family of closed convex subsets of $E \times \mathbf{R}^n$ such that, for each $(u, x) \in E \times \mathbf{R}^n$,

$$H(u, x) = \{(x + Kv, x - \Gamma u) : \\ v \in E, v(t) \in L(t, u(t)) \text{ for } t \in J\}. \tag{4.10}$$

This mapping $H$, in fact, has the property (4.6.1) relative to the mapping

$$h_0 : (u, x) \to (x + KFu, x - \Gamma u) \tag{4.11}$$

of $E \times \mathbf{R}^n$ into itself which arises naturally in the decomposition (4.3) of the mapping $\Psi$ as defined in (4.4).

Thus, by the theorem (4.6), the equation (4.5) will have a solution for every $z \in \mathbf{R}^n$, provided $\Psi$ is injective, and $H$ and $h_0$ satisfy the remaining assumptions in (4.6).

The preceding considerations yield, among others, the following special case of much more general results due to Lasota and Opial [39]:

(4.12) *Let $\lambda$ be a continuous positive valued function in $J = [0, p] \subset \mathbf{R}_+$ and suppose $f$: $J \times \mathbf{R}^n$ is a continuous mapping such that*

$$|f(t, x) - f(t, y)| \leqslant \lambda(t)|x - y| \tag{4.12.1}$$

*for every $t \in J$ and every $x, y$ in $\mathbf{R}^n$.*

*Let $M$, $N$ be endomorphisms of $\mathbf{R}^n$ such that $M + N$ is nonsingular, and put $m = |(M + N)^{-1}N|$.*

*If*

$$\int_0^p \lambda(t)\, dt < \ln\left(1 + \frac{1}{m}\right) \tag{4.12.2}$$

*then, given any $z \in \mathbf{R}^n$, there is a unique solution $v$ in $J$ of the boundary value problem*

$$\dot{v}(t) = f(t, v(t)), \quad Mv(0) + Nv(p) = z. \tag{4.12.3}$$

Further results may be found in [5] (cf., also, [32]).

## III. BROUWER'S FIXED POINT THEOREM AND SOME OF ITS APPLICATIONS

**1.** The famous fixed point theorem of Brouwer plays a central role in the topology of Euclidean space [10].

(1.1) *Every continuous mapping of a closed ball in $\mathbf{R}^n$ into itself has at least one fixed point.*

A proof using little homology theory may be found in [25], for other proofs of (1.1), and of theorems equivalent to (1.1), see [23], [24], for example.

It is clear at once that all applications of (1.1) to differential equations will require geometric considerations of the qualitative behavior of solutions in the underlying coordinate space rather than more analytic considerations of their properties as members of a particular function space.

The classic example again is the problem of the existence of periodic solutions of a differential equation

$$\dot{x} = A(t)x + f(t, x) \tag{1.2}$$

under a variety of assumptions on the linear and nonlinear parts.

**2.** The following theorem is a special case of far more general results due, in their original form, to Poincaré [1] (cf., for example, [27], [32]):

(2.1) *Let $A(t)$, for every $t \in \mathbf{R}$, be an endomorphism of $\mathbf{R}^n$ such that $t \mapsto A(t)$ is continuous and of period $p(>0)$, and such that all characteristic exponents of the linear differential equation*

$$\dot{x} = A(t)x \tag{2.1.1}$$

*have absolute value less than one.*

*Let $f: \mathbf{R} \times H_0 \to \mathbf{R}^n$ be a continuous lipschitzian mapping for which $f(t + p, x) = f(t, x)$ at every $(t, x) \in \mathbf{R} \times H_0$, $H_0 \subset \mathbf{R}^n$ being an open ball centered at the origin.*

*There exists a constant $\alpha > 0$ such that, if for some closed ball $\overline{U}_0 \subset H_0$ of radius $r$ centered at the origin*

$$\alpha |f(t, x)| < r \tag{2.1.2}$$

*for every $(t, x) \in \mathbf{R} \times \overline{U}_0$, then there exists at least one solution $v$ of* (1.2) *in* $\mathbf{R}$ *which is of period $p$.*

Note that the assumptions on $f$ are satisfied, for example, if

$$\lim_{|x| \to 0} |f(t, x)|/|x| = 0 \tag{2.2}$$

uniformly in $\mathbf{R}$, or if, for every $(t, x) \in \mathbf{R} \times H_0$,

$$f(t, x) = kg(t, x) \tag{2.3}$$

where $k > 0$ is a sufficiently small parameter.

Note also that, by Floquet's theory, the principal fundamental (matrix) solution $R$ of (2.1.1) in $\mathbf{R}$ admits the representation

$$R(t) = Z(t)e^{tB} \tag{2.4}$$

where $Z(t)$, for each $t \in \mathbf{R}$, is a nonsingular endomorphism of $\mathbf{R}^n$ such that $Z(t + p) = Z(t)$, and where $B$ is an endomorphism of

$\mathbf{R}^n$ whose eigenvalues, by virtue of the assumptions, all have negative real part.

Thus, any solution $v$ of (1.2) in $\mathbf{R}$ with period $p$ gives rise to the solution $w: t \mapsto Z^{-1}(t)v(t)$ of the differential equation

$$\dot{y} = By + Z^{-1}(t)f(t, Z(t)y), \tag{2.5}$$

which also has period $p$, and conversely. Therefore, the assertion in (2.1) will be proved if (2.5) admits a solution in $\mathbf{R}$ with period $p$.

Clearly, each solution of (1.2) and, equivalently, of (2.5) is uniquely determined by the value it takes at 0. Hence, in particular, each solution $u$ of (2.5) that exists in $[0, p]$ maps the point through which it passes at 0, namely $u(0)$, into the point it reaches at $p$, namely $u(p)$. Thus, the set of solutions that exist in $[0, p]$, if it is nonempty, generates a mapping $T$ of a subset of $H_0$ into $H_0$. Indeed, because of the uniqueness, and hence the continuous dependence upon initial conditions, of the solutions of (2.5), $T$ is continuous, and, because of the periodicity of (2.5), any fixed point of $T$ corresponds to the initial value of a solution of (2.5) that has period $p$.

The mapping $T$ is often called the Poincaré mapping, or the translation operator, corresponding to the solutions of (2.5) (cf., for example, [36]).

Let us show that, as a result of the assumptions in (2.1), there is a closed convex set $C \subset H_0$ in which $T$ is defined (and continuous), such that $T(C) \subset C$. The assertion in (2.1) then follows directly from (1.1).

Since each eigenvalue of $B$ in (2.5) has negative real part, there exists a positive definite quadratic from $V(y)$ for which

$$(\text{grad } V(y), By) = -|y|^2 \tag{2.6}$$

where $(y, z)$ denotes the usual scalar product in $\mathbf{R}^n$ [37]. Choose $\alpha > 0$ such that $|\text{grad } V(y)| \leqslant \alpha|y|$ whatever $y \in \mathbf{R}^n$. Thus any solution $u$ of (2.5), defined in some interval $J \subset \mathbf{R}$, satisfies at every $t \in J$

$$V'(u(t)) \leqslant -|u(t)|^2 + \alpha|u(t)|\,|Z^{-1}(t)|\,|f(t, Z(t)u(t))|, \tag{2.7}$$

and (2.12) implies that there is an open ball $W_0$ centered at the origin such that

$$V'(u(t)) < 0 \tag{2.8}$$

at every $t \in J$ for which $u(t) \in \complement W_0$. This shows that the set

$$C = \{y \in H_0 \colon V(y) \leqslant c\}, \tag{2.9}$$

for a suitably chosen constant $c > 0$, contains with each point $y \in C$, in particular, the value $u(p)$ of the solution $u$ of (2.5) that equals $y$ at 0. In other words, the mapping $T$ is well defined in $C$ and, in fact, $T$ maps $C$ into itself.

This method of proof is very much in the spirit of the so-called Second Method of Lyapunov which plays an important role in stability theory (cf., for example, [38], [40]). The latter's use as a technique in proving results such as (2.1) originated with Lefschetz [41] (cf., also, [40]).

**3.** Another illustration of the use of Brouwer's theorem (1.1) concerns the broad class of second order (scalar) differential equations of the form

$$\ddot{x} + f(x, \dot{x})\dot{x} + g(x) = e(t) \tag{3.1}$$

which arise in many engineering applications.

(3.2) *Let $e, g, f$ be sufficiently smooth continuous real valued functions defined in* $\mathbf{R}$ *and* $\mathbf{R}^2$, *respectively, so that the solutions of* (3.1) *are uniquely determined by the initial conditions.*

*Suppose there are positive constants $r$, $m$, $M$ and $p$ such that*

$$\begin{aligned} \text{(i)}\quad & f(x, y) > m > 0 \quad \textit{for} \quad |x| > a, |y| > a, \\ & f(x, y) > -M \quad \textit{otherwise}; \end{aligned} \tag{3.2.1}$$

*and*

$$\text{(ii)}\quad xg(x) > 0 \quad \textit{for} \quad |x| > a$$

*and*

$$\lim_{|x| \to +\infty} |g(x)| = +\infty,$$
$$\lim_{|x| \to +\infty} g(x) \Big/ \int_0^x g(s)\, ds = 0; \tag{3.2.2}$$

*and*

(iii) $e(t + p) = e(t)$ *for every* $t \in \mathbf{R}$.

*Then there exists at least one solution of* (3.1) *in* $\mathbf{R}$ *which has period* $p$.

This result is due to Levinson [42]. Its proof depends on geometric considerations in the phase plane, writing (3.1) as the system

$$\dot{x} = y$$
$$\dot{y} = -f(x, y)y - g(x) + e(t). \tag{3.3}$$

More precisely, Levinson explicitly constructs a closed set in the plane which is bounded by finitely many Jordan arcs, contains the origin as interior point, and has the following property: any orbit corresponding to a solution of (3.3) that starts from an interior point remains in the interior forever after. Thus that set is mapped into itself by the Poincaré mapping, associated with the solutions of (3.3), and the assertion follows directly from (1.1).

Similar geometric constructions have been used widely, for second order scalar differential equations as well as for two-dimensional systems of first order [48], [49] (cf., for example, [40]).

**4.** A simple general result in the same spirit, essentially, but based on somewhat more analytic considerations, relates the existence of a periodic solution of a differential equation of the form

$$\dot{x} = f(t, x) \tag{4.1}$$

to the existence of a bounded solution [9].

(4.2) *Suppose* $f\colon \mathbf{R} \times W \to \mathbf{R}^n$ *is a continuous mapping for which there is a constant* $p > 0$ *such that* $f(t + p, x) = f(t, x)$ *at every* $(t, x) \in \mathbf{R} \times W$, *W being an arbitrary subset of* $\mathbf{R}^n$.

*There exists a solution of* (4.1) *in* $\mathbf{R}$ *with period p if and only if there exists a solution* $v$ *of* (4.1) *in* $\mathbf{R}_+$ *such that* (i) $\overline{v(\mathbf{R}_+)}$ *is compact and contained in W and* (ii) *there is a divergent sequence of positive integers*, $(n_k)$, *for which*

$$\lim_{k \to \infty} |v((n_k + 1)p) - v(n_k p)| = 0. \tag{4.2.1}$$

Observe that the first condition holds, in particular, if $W$ is a closed set and $v$ is a bounded solution of (4.1) in $\mathbf{R}_+$.

The necessity of the assumptions is quite obvious. The sufficiency part follows from the fact that, for any integer $n_k \geqslant 1$, the mapping $v_k : t \mapsto v(t + n_k p)$ is also a solution of (4.1) in $\mathbf{R}_+$. Indeed, since the values of each $v_k$ belong to $\overline{v(\mathbf{R}_+)}$, the set $\{\hat{v}_k\}$ of restrictions of $v_k$ to $[0, p] \subset \mathbf{R}_+$ is equicontinuous and, by Ascoli's theorem, relatively compact in the Banach space of continuous mappings of $[0, p]$ into $\mathbf{R}^n$. Thus, there is a continuous mapping $\hat{v} : [0, p] \to \mathbf{R}^n$ which is a solution of (4.1) in $[0, p]$, and (4.2.1) implies that

$$\hat{v}(0) = \lim v_{k_i}(0) = \hat{v}(p).$$

Therefore, $\hat{v}$ is in fact a solution of (4.1) in $\mathbf{R}$ which has period $p$.

Related comparison-type results, and further references, may be found in [9].

## IV. SOME FURTHER RESULTS AND NEW DIRECTIONS

**1.** The classic fixed point theorems of Brouwer, Schauder and Banach have been generalized and extended in countless directions. The brief remarks that follow are intended to offer but a glimpse on the wealth of research that has been, and continues to be done in this vast area.

The requirement in Banach's theorem (I. 1.1) that the mapping be a strict contraction is obviously very restrictive. Efforts to remove it completely were doomed because there are simple contractions (of a closed unit ball into itself) which admit no fixed point at all. The following result is essentially the best possible [12] (cf., also, [25] [34]):

(1.1) *Let $E$ be a uniformly convex Banach space* or a Hilbert space, let $C \subset E$ be a bounded closed convex set, and suppose $T : C \to E$ is a contraction, i.e.,*

$$|Tu_1 - Tu_2| \leqslant |u_1 - u_2| \tag{1.1.1}$$

*for any $u_1$, $u_2$ in $C$. If $T(C) \subset C$, then there exists at least one point $v \in C$ such that $Tv = v$.*

It is worth noting that the assumptions imply, for any sequence $(u_n)$ of points of $C$, that if $(u_n)$ converges weakly to a point $u$ and if $(u_n - Tu_n)$ converges strongly to a point $v$, then $u - Tu = v$. This simple property is often of crucial importance.

A very useful and convenient variant of Schauder's theorem (II. 1.1) is the theorem of Rothe which can be stated as follows (cf., for example, [35]):

(1.2) *Let $E$ be a Banach space, let $\overline{U} \subset E$ be a closed ball centered at the origin, and suppose $T : \overline{U} \to E$ is a continuous mapping such that $T(\overline{U})$ is relatively compact. If $T(\partial \overline{U}) \subset \overline{U}$ then there exists a point $v \in \overline{U}$ such that $Tv = v$.*

For various applications of (1.2), we refer to [13] and [35], for example. In another series of generalizations of Schauder's theorem (II. 1.1), the existence of fixed points is derived from hypotheses on some iterate of the mapping rather than on the mapping itself. These so-called asymptotic fixed point theorems have found many applications in the theory of functional differential equa-

---

*A Banach space $E$ is uniformly convex if, for any sequences $(u_n)$, $(v_n)$ with $|u_n| \leqslant 1$, $|v_n| \leqslant 1$, $\lim|u_n + v_n| = 2$ implies $\lim|u_n - v_n| = 0$.

tions, which also provided the original impetus for them (cf., for example, [11], [14], [28]).

(1.3) *Let $E$ be a Banach space and suppose $T : E \to E$ is a continuous mapping such that $T(B)$ is relatively compact for every bounded set $B \subset E$ and there is a bounded set $B_0 \subset E$ and, for each $u \in E$, an integer $j \geqslant 1$ for which $T^j(u) \in B_0$. Then there is a point $v \in B_0$ such that $Tv = v$.*

This result is contained in [33], where references to previous work are given also.

**2.** Krasnoselskii was the first to extend Banach's contraction mapping principle (I. 1.1) to compact perturbations of strict contractions [35]. These mappings belong to the broad class of mappings that have come to be called $\alpha$-contractions, which were introduced by Darbo at about the same time [23]. The term derives from the measure of non-compactness due to Kuratowski [37] which is but one of several that appear to be useful [52].

Kuratowski's measure of non-compactness of a bounded set $B$ in a Banach space $E$ is the greatest lower bound, $\alpha(B)$, of constants $d > 0$ such that $B$ admits a finite covering by sets with diameter not exceeding $d$. Thus, for any bounded set $B \subset E$, $\alpha(B) = \alpha(\bar{B})$ and, more important, $\alpha(B) = 0$ if and only if $B$ is relatively compact. Moreover, for any sequence $(B_n)$ of bounded closed sets $B_n \subset E$ such that $B_{n+1} \subset B_n$ for each $n \geqslant 1$, $\lim \alpha(B_n) = 0$ implies that $\cap B_n$ is nonempty and compact.

A continuous mapping $T$ of a set $U \subset E$ into $E$ is said to be condensing (on $U$, relative to $\alpha$) if, for every bounded set $B \subset U$, $\alpha(T(B)) \geqslant \alpha(B)$ implies $\alpha(B) = 0$. In particular, a continuous mapping $T : U \to E$ is an $\alpha$-contraction if there is a constant $k \in [0, 1[$ such that

$$\alpha(T(B)) \leqslant k\alpha(B) \tag{2.1}$$

for every bounded set $B \subset U$.

The basic fixed point theorem for condensing mappings is due to Sadovskii [51]. It extends Darbo's original theorem [23] which in turn, formally is the extension of (I. 1.1) to mappings satisfying (2.1) (cf., also, [31]).

(2.2) *Let $E$ be a Banach space, let $C \subset E$ be a bounded closed set, and suppose $T$ is a condensing mapping of $C$ into itself. Then there exists a point $v \in C$ such that $Tv = v$.*

The proof rests on the fact that the mapping $u \mapsto (1/n)u_0 + (1-(1/n))Tu$, for any point $u_0 \in C$ and each integer $n \geqslant 1$, is an $\alpha$-contraction on $C$ for the constant $1-(1/n)$, and hence has a fixed point $u_n \in C$, and that the set of these fixed points is relatively compact.

Related results may be found in [52], for example.

Fixed point theorems for condensing mappings and $\alpha$-contractions have proved themselves particularly useful in the theory of functional differential equations and, especially, of neutral functional differential equations where the classical fixed point results fail to apply. This is also true of the recent generalizations to condensing mappings, due to Hale [31] and others (cf., for example, [30]), of such asymptotic fixed point theorems as (1.3) and others.

The following theorem on the existence of a solution of a differential equation

$$\dot{x} = f(t, x) \tag{2.3}$$

in a Banach space deserves mention here because it extends the classical theory (cf., for example (II. 2.2)) beyond the cases requiring uniform continuity [16]:

(2.4) *Let $J \subset \mathbf{R}$ be a compact interval, let $H$ be an open subset of a Banach space $X$, and suppose $f$ is a continuous mapping of $J \times H$ into $X$.*

*If, given $(a, y) \in J \times H$, there is a closed ball $\overline{W} \subset H$ with center $y$ such that $f$ is bounded in $J \times \overline{W}$ and that*

$$\int_{0+} \frac{dr}{rL(r)} = +\infty \tag{2.4.1}$$

*where L is defined by setting, for any* $r > 0$,

$$L(r) = \sup\left\{\alpha(f(J \times B))/\alpha(B) : B \subset \overline{W}, \alpha(B) \geqslant r\right\}, \quad (2.4.2)$$

*then there exists a compact interval* $J_a \subset J$ *containing a in which* (2.3) *has at least one solution* $v$ *with* $v(a) = y$.

Observe that (2.4.1) holds, in particular, if $f$ has the property that, for some $k > 0$,

$$\alpha(f(J \times B)) \leqslant k\alpha(B) \quad (2.5)$$

whatever the set $B \subset \overline{W}$.

The proof of (2.4) rests upon considerations of the mapping (II. 2.3), as does the proof of (II. 2.2), and upon an ingenious construction of a nested sequence of closed convex sets whose common intersection is nonempty, compact, and invariant for that mapping. Thus, this construction, in essence, yields the compactness that follows from Ascoli's theorem in the finite dimensional case.

For an entirely different approach to existence theorems see, for example, [43] where further references are given.

An extensive summary of recent research on the solution of the general equation

$$v + KFv = w, \quad (2.6)$$

which underlies our entire discussion, may be found in [15].

## REFERENCES

1. H. A. Antosiewicz, "Forced periodic solutions of systems of differential equations," *Ann. of Math.*, **57** (1953), 314–317; *ibid.*, **58** (1953), 592.

2. ———, "On the existence of periodic solutions of nonlinear differential equations," *Proc. Colloques Internationaux C.N.R.S.*, **148** (1965), 213–216.

3. ———, "Boundary value problems for nonlinear ordinary differential equations," *Pacific J. Math.*, **17** (1966), 191–197.

4. ———, "Un analogue du principe du point fixe de Banach," *Ann. Mat. Pura Appl.*, **74** (1966), 61–64.

5. ———, "Linear problems for nonlinear ordinary differential equations," *Proc. U.S.-Japan Sem. Diff. and Funct. Eqns.*, W. A. Benjamin, New York, 1967, 1–11.

6. ———, "A fixed point theorem and the existence of periodic solutions," *Proc. 5th Internat. Conf. Nonlin. Oscill., Kiev,* 1969, 40–44.

7. S. Banach, "Sur les opérations dans les ensembles abstraits et leur application aux équations intégrales," *Fund. Math.*, **3** (1922), 133–181.

8. S. Bancoft, J. K. Hale, and D. Sweet, "Alternative problems for nonlinear functional equations," *J. Differential Equations*, **4** (1968), 40–56.

9. R. Becker and G. Vidossich, "Some applications of a simple criterion for the existence of periodic solutions of ordinary differential equations," *J. Math. Anal. Appl.*, **48** (1975), 51–60.

10. L. E. J. Brouwer, "Uber eineindeutige, stetige Transformationen von Flächen in sich," *Math. Ann.*, **69** (1910), 176–180.

11. F. E. Browder, "On a generalization of the Schauder fixed point theorem," *Duke Math. J.*, **26** (1959), 291–303.

12. ———, "Non-expansive nonlinear operators in a Banach space," *Proc. Nat. Acad. Sci. U.S.A.*, **54** (1965), 1041–1044.

13. ———, "Nonlinear equations of evolution and nonlinear accretive operators in a Banach space," *Bull. Amer. Math. Soc.*, **73** (1967), 867–874.

14. ———, "Asymptotic fixed point theorems," *Math. Ann.*, **185** (1970), 38–60.

15. ———, "Nonlinear functional analysis and nonlinear integral equations of Hammerstein and Urysohn type," *Contrib. Nonlinear Functional Analysis*, Academic Press, New York, 1971.

16. A. Cellina, "On the existence of solutions of ordinary differential equations in Banach spaces," *Funkcial. Ekvac.*, **14** (1971), 129–136.

17. ———, "On the nonexistence of solutions of differential equations in non-reflexive spaces," *Bull. Amer. Math. Soc.*, **78** (1972), 1069–1072.

18. L. Cesari, "Functional analysis and periodic solutions of nonlinear differential equations," *Contrib. Differential Equations*, **1** (1963), 149–187.

19. S. H. Chang, "Periodic solutions of certain second order nonlinear differential equations," *J. Math. Anal. Appl.*, **49** (1975), 263–266.

20. R. Conti, "Recent trends in the theory of boundary value problems for ordinary differential equations," *Boll. Un. Mat. Ital.*, **22** (1967), 135–178.

21. W. A. Coppel, *Stability and Asymptotic Behavior of Differential Equations*, Heath, Boston, 1965.

22. J. Cronin, "Equations with bounded nonlinearities," *J. Differential Equations*, **14** (1973), 581–596.

23. G. Darbo, "Punti uniti in trasformazioni a condominio non compatto," *Rend. Sem. Mat. Univ. Padov*, **24** (1955), 84–92.

24. J. Dugundji, *Topology*, Allyn and Bacon, Boston, 1966.

25. N. Dunford and J. T. Schwartz, *Linear Operators, Part I: General Theory*, Interscience, New York, 1957.

26. L. M. Graves, *The Theory of Functions of Real Variables*, McGraw-Hill, New York, 1946.

27. J. K. Hale, *Oscillations in Nonlinear Systems*, McGraw-Hill, New York, 1963.

28. ———, "Functional differential equations," *Appl. Math. Sci.*, vol. 3, Springer-Verlag, 1971.

29. ———, "Continuous dependence of fixed points of condensing maps," *J. Math. Anal. Appl.*, **46** (1974), 388–394.

30. J. K. Hale and O. Lopes, "Fixed point theorems and dissipative processes," *J. Differential Equations*, **13** (1973), 391–402.

31. J. K. Hale, $\alpha$ contractions and differential equations, *Équations Différentielles et Fonctionnelles Nonlinéaires* (P. Janssens, J. Mawhin, N. Rouche, eds.), Hermann, Paris, 1973, 15–41.

32. P. Hartman, *Ordinary Differential Equations*. Wiley, New York, 1964.

33 W. A. Horn, "Some fixed point theorems for compact mappings and flows on a Banach space," *Trans. Amer. Math. Soc.*, **149** (1970), 391–404.

34. W. A. Kirk, "A fixed point theorem for mappings which do not increase distances," *Amer. Math. Monthly*, **72** (1965), 1004–1006.

35. M. A. Krasnoselskii, *Topological Methods in the Theory of Nonlinear Integral Equations*, Pergamon Press, Oxford, 1964.

36. ———, "The operator of translation along the trajectories of differential equations," *Amer. Math. Soc. Transl. Math. Monographs*, vol. **19**, Providence, 1968.

37. C. Kuratowski, *Topologie*, 4th ed., Warsaw, 1962.

38. J. Lasalle and S. Lefschetz, *Stability by Liapunov's Direct Method, with Applications*, Academic Press, New York, 1961.

39. A. Lasota and Z. Opial, "On the existence of solutions of linear problems for ordinary differential equations," *Bull. Acad. Polon. Sci. Sér. Sci. Math. Astronom. Phys.*, **14** (1966), 371–375.

40. S. Lefschetz, *Differential Equations: Geometric Theory*, 2nd ed., Wiley, New York, 1963.

41. ———, "Existence of periodic solutions for certain differential equations," *Proc. Nat. Acad. Sci. U.S.A.*, **29** (1943), 29–32.

42. N. Levinson, "On the existence of periodic solutions for second order differential equations with a forcing term," *J. Math. Phys.*, **22** (1943), 41–48.

43. R. H. Martin, Jr., "Differential equations on closed subsets of a Banach space," *Trans. Amer. Math. Soc.*, **179** (1973), 399–414.

44. J. L. Massera, "Sur l'existence de solutions bornées et périodiques des systèmes quasi-linéaires d'équations différentielles," *Ann. Mat. Pura Appl.*, **51** (1960), 95–105.

45. J. L. Massera and J. J. Schäffer, *Linear Differential Equations and Function Spaces*, Academic Press, New York, 1966.

46. J. Mawhin, "Periodic solutions of nonlinear functional differential equations" *J. Differential Equations*, **10** (1971), 240–261.

47. W. V. Petryshyn, "On the approximation-solvability of equations involving $A$-proper and pseudo-$A$-proper mappings," *Bull. Amer. Math. Soc.*, **81** (1975), 223–312.

48. G. E. H. Reuter, "A boundedness theorem for non-linear differential equations of the second order, I," *Proc. Cambridge Philos. Soc.*, **47** (1951), 49–54.

49. ———, "Boundedness theorems for nonlinear differential equations of the second order, II," *J. London Math. Soc.*, **27** (1952), 48–58.

50. N. Rouche and J. Mawhin, *Equations Différentielles Ordinaires, I, II*, Masson et Cie., Paris, 1973.

51. B. N. Sadovskii, "On a fixed point theorem," *Funkcional. Anal. i Priložen*, **1** (1967), 74–76.

52. ———, "Limit compact and condensing operators," *Uspehi Mat. Nauk*, **27** (163)(1972), 81–146.

53. J. Schauder, "Der Fixpunktsatz in Funktionalräumen," *Studia Math.*, **2** (1930), 171–180.

54. H. Tychonoff, "Ein Fixpunktsatz," *Math. Ann.*, **111** (1935), 767–776.

55. J. A. Yorke, "A continuous differential equation in Hilbert space without existence," *Funkcial. Ekvac.*, **13** (1970), 19–21.

# THE ALTERNATIVE METHOD IN NONLINEAR OSCILLATIONS

*Lamberto Cesari*

## 1. AN ALTERNATIVE SCHEME

The injection of methods of functional analysis (Cesari [3, 4], 1963) in the classical bifurcation process of Poincaré [62], Lyapunov [43], and Schmidt [64], and extensive subsequent work, have made this process a fine tool in nonlinear analysis, particularly in the difficult problems "at resonance" in the usual terminology. The general theory which has ensued, with all its variants and ramifications, is often referred to as bifurcation theory, or the alternative method.

To be specific, let us consider an equation of the form

$$Ex = Nx, \tag{1}$$

where $E : \mathfrak{D}(E) \rightarrow Y$ is a linear operator, and $N : \mathfrak{D}(N) \rightarrow Y$ is an operator nonnecessarily linear, both $E$ and $N$ with domains

---

This research was partially supported by AFOSR Research Project 71-2122 at the University of Michigan.

$\mathfrak{D}(E)$, $\mathfrak{D}(N)$ in a Banach space $X$ and ranges $\mathfrak{R}(E)$, $\mathfrak{R}(N)$ in a Banach space $Y$, $\mathfrak{D}(E) \cap \mathfrak{D}(N) \neq \emptyset$.

Let us assume for a moment that the following holds (we shall see in Section 3 that these assumptions are quite natural and easily verified in the so-called selfadjoint case). Let us assume that there are projection operators $P : X \to X$, $Q : Y \to Y$ (that is, linear, bounded, idempotent, or $PP = P$), and a linear operator $H$, which we shall consider as a partial inverse of $E$, satisfying

$$(\mathrm{k}_1)\quad H(I - Q)Ex = (I - P)x \qquad \text{for all } x \in \mathfrak{D}(E),$$

$$(\mathrm{k}_2)\quad QEx = EPx \qquad \text{for all } x \in \mathfrak{D}(E),$$

$$(\mathrm{k}_3)\quad EH(I - Q)Nx = (I - Q)Nx \qquad \text{for all } x \in \mathfrak{D}(E) \cap \mathfrak{D}(N).$$

Under these assumptions, the equation $Ex = Nx$ is equivalent to the system of two equations

$$x = Px + H(I - Q)Nx, \tag{2}$$

$$Q(Ex - Nx) = 0. \tag{3}$$

Indeed, if (1) is satisfied, then by applying $Q$ to equation (1) we obtain (3). By applying $H(I - Q)$ to (1) we obtain $H(I - Q)Ex = H(I - Q)Nx$, and by using $(\mathrm{k}_1)$ we obtain (2). Conversely, if (2) and (3) are satisfied, then, by applying $E$ to (2), we have $Ex - EPx = EH(I - Q)Nx$, and by using $(\mathrm{k}_2)$, $(\mathrm{k}_3)$, also $Ex - QEx = (I - Q)Nx$, or $Ex - Nx = Q(Ex - Nx) = 0$. Equations (2) and (3) are usually denoted as the auxiliary and the bifurcation equations, respectively.

Let $X_0 = PX$, $X_1 = (I - P)X$, $Y_0 = QY$, $Y_1 = (I - Q)Y$, so that $X$ and $Y$ have the decompositions $X = X_0 + X_1$, $Y = Y_0 + Y_1$ (direct sums), $X_0 \cap X_1 = \{0\}$, $Y_0 \cap Y_1 = \{0\}$. The linear subspace of $X$ of all elements $x \in X$ for which $Ex = 0$ is often denoted as the null space of $E$, or kernel of $E$, and denoted by $\ker E$. When there are elements $x \neq 0$ in $X$ with $Ex = 0$, that is, $\ker E$ is not trivial, then we say that we have a resonance case. We shall always assume that $\ker E \subset X_0$. Also, we shall assume that $Y_1 \subset \mathfrak{R}(E)$, $\mathfrak{D}(H) = Y_1$, $\mathfrak{R}(H) = \mathfrak{D}(E) \cap X_1$. Thus, $EHy = y$ for all $y \in Y_1$, $HEx = x$ for all $x \in \mathfrak{D}(E) \cap X_1$. Often we may assume $\ker E = X_0$, $Y_1 = \mathfrak{R}(E)$. In this particular situation the bifurcation equation reduces to $QNx = 0$.

If we write $x^* = Px$, then the auxiliary equation can always be written in the form of a fixed point statement:

$$x = T_1 x, \quad T_1 x = x^* + H(I - Q)Nx, \tag{4}$$

where $T_1$ is easily seen to map the fiber $P^{-1}x^*$ into itself. The auxiliary equation can also be written in the form of a Hammerstein equation:

$$x + KNx = x^*, \quad K = -H(I - Q). \tag{5}$$

The scheme above was proposed by Cesari in 1963 [3, 4] for the "selfadjoint" case (briefly, $X = Y$, $P = Q$), and soon extended to "nonselfadjoint" cases by Locker [42], Hale, Bancroft, and Sweet [26], Hale [23]. (See Hale [25].)

## 2. REDUCTION TO AN ALTERNATIVE PROBLEM BY CONTRACTION MAPS

There are certain well-known situations where the auxiliary equation (2) is uniquely solvable for every $x^* \in X_0$, and we may write its solution in the form $x = \mathfrak{T}(x^*)$ with $Px = P\mathfrak{T}(x^*) = x^*$, $x^* \in X_0$, or

$$x = [I - H(I - Q)N]^{-1}x^*.$$

By replacing $x = \mathfrak{T}(x^*)$ in the bifurcation equation (3), we have reduced the original problem $Ex = Nx$, to the sole equation:

$$Q(E - N)\mathfrak{T}(x^*) = 0, \text{ or} \tag{6}$$

$$T_2 x^* = Q(E - N)[I - H(I - Q)N]^{-1}x^* = 0, \tag{6'}$$

or other analogous forms, all being equations in $X_0$. This is often a finite dimensional space. We say that the original problem $Ex = Nx$ has been reduced to an alternative one.

Cesari proved [3, 4] the general statement that, for selfadjoint problems (see Section 3) and $N$ uniformly Lipschitzian in a suitable domain, it is always possible to choose $P = Q$ and $H$ in such a way that $T_1$ is a contraction map on the fiber $P^{-1}x^*$ of $x^*$ with respect to $P$, and thus $T_1$ has a fixed point $\mathfrak{T}(x^*)$ on $P^{-1}x^*$ by Banach's fixed point theorem for every $x^* \in X_0$. In other words, equation (1) admits of an alternative problem. This statement has been extended to nonselfadjoint problems by Hale [23] under the assumptions that $N$ is uniformly Lipschitzian in a suitable domain and that $X$ is a Hilbert space.

We refer to [7, 8, 9, 23, 25] for recent expositions and applications, and in particular to [6] for an application to a nonlinear Dirichlet problem in the plane. We mention here that, for what concerns ordinary differential equations of the perturbation type, existence theorems of periodic solutions, which had been proved by Cesari and Hale by a variant of the above process for selfadjoint problems, have been extended by Nagle [56] to general nonselfadjoint boundary value problems within the alternative scheme just mentioned (see Section 5). Analogously, nonselfadjoint elliptic nonlinear problems of the perturbation type have been discussed by Shaw [65] in terms of Sobolev spaces and the alternative method (see Section 4). The nonlinear perturbation type wave equation was discussed by Cesari [5] and Hale [24] by the alternative method, and in this situation the kernel of $E$, and therefore the alternative problem, are infinite dimensional.

## 3. THE SELFADJOINT CASE

Let us assume here that $X$ is a real separable Hilbert space with inner product $(x, y)$ and norm $\|x\| = (x, x)^{1/2}$. (In applications, $E$ may be a linear differential operator on some domain $G \subset E^\nu$, $\nu \geq 1$, with given homogeneous linear boundary conditions on the boundary $\partial G$ of $G$.) Here we need to know that the associated linear problem $Ex + \lambda x = 0$ has countably many real eigenvalues $\lambda_i$ and eigenelements $\phi_i$, $\lambda_i \leq \lambda_{i+1}$, $E\phi_i + \lambda_i\phi_i = 0$, $i = 1, 2, \ldots,$ with $\lambda_i \to +\infty$ as $i \to \infty$, and $\{\phi_i, i = 1, 2, \ldots\}$ is a complete orthonormal system in $X$. (Then, in the situation just mentioned, we would have $X = L_2(G)$.)

Thus, every element $x \in X$ has Fourier series $x = \Sigma_1^\infty c_i \phi_i$, $c_i = (x, \phi_i)$, which is convergent in $X$. If $d = \Sigma_1^\infty d_i \phi_i$ is any other element, then $(x, y) = \Sigma_1^\infty c_i d_i$, $\|x\| = (\Sigma_1^\infty c_i^2)^{1/2}$. Since $\lambda_i \leqslant \lambda_{i+1}$ and $\lambda_i \to +\infty$, there is some integer $m$ so that $\lambda_{m+1} > 0$, and we shall define $P : X \to X$ by taking $Px = \Sigma_1^m c_i \phi_i$, $c_i = (x, \phi_i)$, $i = 1, \ldots, m$. Then, the range $X_0 = PX$ of $P$ is the $m$-dimensional space $X_0 = \{\phi_1, \ldots, \phi_m\}$ spanned by $\phi_1, \ldots, \phi_m$. If $X_1$ is the complementary space, then any element $x \in X_1$ has Fourier series $x = \Sigma_{m+1}^\infty c_i \phi_i$, and we can define $H : X_1 \to X_1$ by taking

$$Hx = -\sum_{m+1}^{\infty} c_i \lambda_i^{-1} \phi_i, \quad x \in X_1.$$

This is possible since we chose $m$ so that $\lambda_{m+1} > 0$ and thus $\lambda_i \geqslant \lambda_{m+1} > 0$ for all $i \geqslant m+1$. Then we also have

$$\|Hx\| = \left( \sum_{m+1}^{\infty} c_i^2 \lambda_i^{-2} \right)^{1/2} \leqslant \lambda_{m+1}^{-1} \left( \sum_{m+1}^{\infty} c_i^2 \right)^{1/2} = \lambda_{m+1}^{-1} + \|x\|,$$

or $\|Hx\| \leqslant k\|x\|$, $k = \lambda_{m+1}^{-1}$, for all $x \in X_1$. We see that $H$ is a linear bounded operator, and that $H$ can always be made into a contraction by taking $m$ so that $\lambda_{m+1} > 1$.

It is apparent that $P : X \to X$ is a linear projection operator, $PP = P$, $\|P\| = 1$, and if $I$ denotes the identity operator in $X$, then also $I - P$ is a projection operator, $\|I - P\| = 1$, and $X_1 = (I - P)X$. If $\mathfrak{D}(E)$ denotes the linear subspace of $X$ of all $x = \Sigma_1^\infty c_i \phi_i$, $c_i = (x, \phi_i)$, with $\Sigma_1^\infty c_i^2 \lambda_i^2 < \infty$, then $E : \mathfrak{D}(E) \to X$, and $P, E, H$ have the properties $(k_{123})$ of Section 1 with $X = Y$, $P = Q$, that is, $(h_1)$ $H(I - P)E = I - P$; $(h_2)$ $PE = EP$; $(h_3)$ $EH(I - P) = I - P$.

Thus, if $N : X \to X$ is any operator in $X$, then the equation $Ex = Nx$ in $X$ is equivalent to the system of auxiliary and bifurcation equations

$$x = x^* + H(I - P)Nx, \quad Px = x^* \in X_0,$$
$$P(Ex - Nx) = 0,$$

and the auxiliary equation is of the form $x = Tx$ with $T$ defined by $Tx = x^* + H(I - P)Nx$. Note that, if $N : X \to X$ is uniformly Lipschitzian in $X$ of given constant $L$, that is, $\|Nx_1 - Nx_2\| \leqslant L\|x_1 - x_2\|$ for all $x_1, x_2 \in X$, then by taking $m$ so large that $\lambda_{m+1} > L$, the map $T$ is a contraction on $P^{-1}x^*$, or $x^* + X_1$, of constant $kL = \lambda_{m+1}^{-1}L < 1$. Thus, for every $x^* \in X_0$, the auxiliary equation has a unique solution $x = Tx = \mathfrak{T}x^*$, as mentioned in Section 2, and the bifurcation equation takes the form $P(E - N)\mathfrak{T}x^* = 0$, an equation in $X_0$, a finite dimensional space (the alternative problem) [3, 4].

If $x^*$ is written in the form $x^* = \gamma_1\phi_1 + \cdots + \gamma_m\phi_m$, then we have

$$P(E - N)\mathfrak{T}x^* = c_1(x^*)\phi_1 + \cdots + c_m(x^*)\phi_m,$$

and the bifurcation equation $P(E - N)\mathfrak{T}x^* = 0$ takes the form

$$\text{(B)} \qquad c_1(x^*) = 0, \ldots, c_m(x^*) = 0,$$

a system of $m$ equations in the $m$ unknowns $\gamma_1, \ldots, \gamma_m$.

A qualitative analysis has been the goal of the entire research. However, numerical results have shown to be of interest (see, e.g., Section 4). Also, Sanchez [63] has devised a method of successive approximations for the solution of the auxiliary and bifurcation equations and therefore of the given equation $Ex = Nx$.

## 4. ERROR BOUND ESTIMATES

For $E, N, X, P, H$ as before, an element $x_0^* \in X_0$, $x_0^* = \gamma_1\phi_1 + \cdots + \gamma_m\phi_m$, is said to be a Galerkin approximation to a (possible) solution $x \in S$ of the equation $Ex = Nx$, provided

$$P(E - N)x^* = 0.$$

For $P(E - N)x^* = \Gamma_1(x^*)\phi_1 + \cdots + \Gamma_m(x^*)\phi_m$, this equation takes the form

$$\text{(G)} \qquad \Gamma_1(x^*) = 0, \ldots, \Gamma_m(x^*) = 0,$$

also a system of $m$ equations in the $m$ unknowns $\gamma_1, \ldots, \gamma_m$.

The difference with (B) is that, while these equations (G) give merely an approximate solution $x_0^*$ to the equation $Ex = Nx$, the bifurcation equations (B) yield an element $x^*$ of $X_0$ such that $x = \mathfrak{T}x^*$ is an exact solution of $Ex = Nx$.

The considerations of Section 3 may lead to an answer to the following two questions: 1. Given any Galerkin approximation $x_0^* \in X_0$, is there an exact solution $x \in X$ in some neighborhood $W$ of $x_0^*$ in $X$? 2. Is it possible to estimate the error $\|x - x_0^*\|$?

To see this briefly, let us first compute $\Delta x_0^* = Ex_0^* - Nx_0^*$, that is, the error in the equation $Ex - Nx = 0$, and then let us compute $\delta x_0^* = H(I - P)\Delta x_0^*$, and $\rho = \|\delta x_0^*\|$. Then, we have $Ex_0^* = Nx_0^* + \Delta x_0^*$, and by applying $H(I - P)$ and using $(h_1)$, also

$$(I - P)x_0^* = H(I - P)Ex_0^* = H(I - P)Nx_0^* + \delta x_0^*,$$

where $x_0^* \in X_0$, hence $(I - P)x_0^* = 0$, and $H(I - P)Nx_0^* + \delta x_0^* = 0$.

Now let $c, d$, $0 < c < d$, denote any two real numbers, and let $W_0$, $W$ denote the neighborhoods of $x_0^*$ in $X_0$ and $X$ defined by $W_0 = [x \in X_0 | \|x - x_0^*\| \leqslant c]$, $W = [x \in X | \|x - x_0^*\| \leqslant d]$. For any $x^* \in W_0$ let $V(x^*)$ denote the set $V(x^*) = [x \in X | Px = x^*, \|x - x_0^*\| \leqslant d]$, that is, the part of the fiber $P^{-1}x^* = x^* + X_1$ contained in $W$. Let us assume that $N$ is defined in $W$ and is there Lipschitzian of constant $L$, that is, $\|Nx_1 - Nx_2\| \leqslant L\|x_1 - x_2\|$ for any two $x_1, x_2 \in W$. Also, let us assume that $kL = \lambda_{m+1}^{-1}L < 1$, and that $d \geqslant (1 - kL)^{-1}(c + \rho)$, that is, $c + kLd + \rho \leqslant d$. Then, for every $x^* \in W_0$, the map $T$ maps $V(x^*)$ into itself and is a contraction there, and has, therefore, a unique fixed point $x = Tx = \mathfrak{T}x^*$ in $V(x^*)$. Indeed, for $x \in V(x^*)$, we have

$$Tx = x^* + H(I - P)Nx,$$

$$PTx = Px^* + PH(I - P)Nx = x^*,$$

$$\begin{aligned} Tx - x_0^* &= x^* - x_0^* + H(I - P)Nx \\ &= x^* - x_0^* + H(I - P)Nx \\ &\quad - H(I - P)Nx_0^* - \delta x_0^*, \end{aligned}$$

$$\begin{aligned} \|Tx - x_0^*\| &\leqslant \|x^* - x_0^*\| + \|H(I - P)(Nx - Nx_0^*)\| + \|\delta x_0^*\| \\ &\leqslant c + kLd + \rho \leqslant d. \end{aligned}$$

This proves that $T : V(x^*) \to V(x^*)$, and we know already from Section 2 that $T$ is a contraction.

Thus, the auxiliary equation, or $x = Tx$, is solvable for every $x^* \in W_0$ with a unique solution $x = Tx = \mathfrak{T}x^*$ in $V(x^*)$, and the problem $Ex - Nx = 0$ is reduced to the bifurcation equation $P(E - N)\mathfrak{T}x^* = 0$.

If we can choose $c$ and $d$ in such a way that this equation has a solution $x^*$ in $W_0$, then we know that $x = Tx = \mathfrak{T}x^* \in W$ is a solution of the given equation $Ex = Nx$. Moreover $d \geqslant \|x - x_0^*\|$ gives an estimate of the error $\|x - x_0^*\|$ between the exact solution $x$ and the Galerkin approximation $x_0^*$ for the equation $Ex = Nx$.

The process sketched here has many variants and can be reworded in Banach spaces and by the use of different norms. It has been applied to a number of particular cases in [3, 4, 6] with $m = 1$ or $m = 2$. Recently, Cesari and Bowman, and Ku, have programmed the process for the use with digital computers and $m = 1$. For instance, the two periodic ordinary differential equations below have been probed for $2\pi$-periodic solutions (harmonics). We list here the first Galerkin approximation $x_0^*$ for some of the constants appearing in the equations, and the corresponding error bound $\varepsilon \geqslant \|x - x_0^*\|_{\text{Sup}}$ in the Sup norm of the difference between $x_0^*$ and the exact solution $x$, which is thereby proved to exist in the $\varepsilon$-neighborhood of $x_0^*$ (in the Sup norm) (Cesari and Bowman [10], I.B.M. 370):

$$x'' + A \sin x + C \sin t = 0,$$

$$A = 1,\ C = 1, \qquad x_0^* = 2.130 \sin t,\ \varepsilon = 0.119;$$

$$A = 1,\ C = 1.5, \qquad x_0^* = 2.496 \sin t,\ \varepsilon = 0.136.$$

$$x'' + Ax^3 + B \sin t = 0,$$

$$A = 0.6,\ B = 0.2, \qquad x_0^* = 0.2038,\ \varepsilon = 0.01;$$

$$A = 1.0,\ B = 0.2, \qquad x_0^* = 0.2067,\ \varepsilon = 0.01;$$

$$A = 1.0,\ B = 0.4, \qquad x_0^* = 0.4862,\ \varepsilon = 0.03.$$

Analogously, the Dirichlet problem $\nabla^2 x = g(\xi, \eta, x)$ in the domain $G = [\xi^2 + \eta^2 < 1]$ with boundary values $x = 0$ on $\partial G$, was probed with $m = 1$ and some functions $g$. These are some of the results with error bounds $\varepsilon \geqslant \|x - x_0^*\|_{\mathrm{Sup}}$ in the Sup norm on $G$ (Ku [36] IBM 370):

$$g = |u| + 1, \quad x_0^* = 0.33493 J_0(2.4048r), \quad \varepsilon = 0.0413;$$
$$g = u^3 + 1, \quad x_0^* = 027914 J_0(2.4048r), \quad \varepsilon = 0.0313,$$

where $r = (\xi^2 + \eta^2)^{1/2}$ and $J_0$ denotes the Bessel function of index zero.

## 5. PROBLEMS OF PERTURBATION TYPE FOR ORDINARY DIFFERENTIAL EQUATIONS

We consider here with Nagle [56] boundary value problems for systems of $n$ first order ordinary differential equations with homogeneous linear boundary conditions, or

$$\begin{gathered} x' - A(t)x = f(t, x, x'), \quad a \leqslant t \leqslant b, \quad x' = dx/dt, \\ B_1 x(a) + B_2 x(b) = 0, \end{gathered} \tag{7}$$

where $x(t) = \mathrm{col}(x_1, \ldots, x_n)$ are the $n$ unknowns, where $A(t) = [a_{ij}(t)]$ is an $n \times n$ matrix with real bounded measurable entries, and $B_1$, $B_2$ are constant $n \times n$ matrices. Thus, in (7) the underlying linear problem is

$$x' - A(t)x = 0, \quad a \leqslant t \leqslant b, \quad B_1 x(a) + B_2 x(b) = 0, \tag{8}$$

or $Ex = 0$, including the boundary conditions in the operator $E$ defined by $Ex = x' - A(t)x$. As usual, a solution $x(t) = \mathrm{col}(x_1, \ldots, x_n)$, $a \leqslant t \leqslant b$, is an $n \times 1$ vector with absolutely continuous (AC) entries in $[a, b]$, satisfying the differential system a.e. in $[a, b]$.

Let $|\ \ |$ denote the Euclidean norm in $E^n$. In (7), $f(t, x, x') = \mathrm{col}(f_1, \ldots, f_m)$ is an $n \times 1$ vector function defined on $[a, b] \times E^{2n}$, whose entries are measurable in $t$ for every $(x, x')$, and continuous in $(x, x')$ for every $t$. Moreover, we assume here that for any given pair of constants $R_1$, $R_2$ there are two other constants $L$, $M$ such that

$$|f(t, x, x')| \leqslant M,$$

$$|f(t, x, x') - f(t, y, y')| \leqslant L[|x - y| + |x' - y'|],$$

for all $a \leqslant t \leqslant b$, $|x|, |y| \leqslant R_1$, $|x'|, |y'| \leqslant R_2$. If $N$ denotes the operator defined in the second member of (7), then problem (7) takes the usual form $Ex = Nx$.

For any $n$-vector $z(t) = (z_1, \ldots, z_n)$, $a \leqslant t \leqslant b$, we denote by $\|z\|_0$ the usual Sup norm, or $\|z\|_0 = \mathrm{Sup}_{a \leqslant t \leqslant b}|z(t)|$, and by $\|z\|_1$ the $L_1$-norm $\|z\|_1 = (b - a)^{-1}\int_a^b |z(t)|\, dt$. Let $X$ denote the (Sobolev) space of all AC vector functions

$$x(t) = (x_1, \ldots, x_n),\ a \leqslant t \leqslant b,$$

for which we may well take the norm $\|x\|_1' = \|x\|_0 + \|x'\|_1$ (since this norm is equivalent to the Sobolev norm $\|x\|_1 + \|x'\|_1$). Let $Y$ be the space of all vector functions $y(t) = (y_1, \ldots, y_n)$ with $L_1$-integrable entries, with norm $\|y\|_1$.

Let $\tilde{A}$ denote the transpose of any given matrix $A$. Then the linear problem adjoint to (8) is

$$d\tilde{y}/dt + \tilde{y}A(t) = 0, \quad \tilde{y}(a) = \tilde{\alpha}B_1, \quad \tilde{y}(b) = -\tilde{\alpha}B_2, \tag{9}$$

or

$$dy/dt + \tilde{A}(t)y = 0, \quad y(a) = \tilde{B}_1\alpha, \quad y(b) = -\tilde{B}_2\alpha, \tag{9$'$}$$

where $y(t) = \mathrm{col}(y_1, \ldots, y_n)$, or $\tilde{y}(t) = \mathrm{row}(y_1, \ldots, y_n)$, and where $\alpha = \mathrm{col}(\alpha_1, \ldots, \alpha_n)$ denotes any arbitrary real $n \times 1$ vector (parametric form of the adjoint problem). Then, the linear operator $E^*$ adjoint to $E$ is defined by relations (9), or (9)$'$, and as usual $\ker E = \mathrm{coker}\, E^*$, $\mathrm{coker}\, E = \ker E^*$. We denote by $p$ and $q$ the dimensions of $\ker E$ and $\mathrm{coker}\, E$, respectively. Fredholm's alternative theorem has now the usual form. Moreover, projection operators $P_0 : X \to X$, $Q_0 : Y \to Y$, mapping $X$ onto $X_0 = P_0(X) = \ker E$, and $Y$ onto $Y_0 = Q(Y) = \mathrm{coker}\, E$, can be defined here by simple algebraic operations (see Hale [25], p. 263; Nagle [56]).

Then problem $Ex = Nx$, can be reduced to a system (2), (3) of auxiliary and bifurcation equations. By an analysis too detailed to be reported here concerning the interplay of $L_1$ and Sup norms, the auxiliary equation can be handled as a Lipschitz map. Thus the same equation is solvable if the Lipschitz constant is $< 1$. For $f$ replaced by $\varepsilon g$ in (7), where $\varepsilon$ is a "small" parameter, this Lipschitz constant can be made $< 1$ by taking $\varepsilon$ sufficiently small. Thus, we are reduced to the bifurcation equation, actually a system of $q$ equations in $p$ unknowns. (See [56].) For $q \leqslant p$ and $g$ smooth, it is enough to verify that the relevant $q \times p$ Jacobian matrix for $\varepsilon = 0$, which is easy to compute, has maximum rank $q$. Then by the implicit function theorem problem (7) has at least a solution.

Theorems are proved in [56] which extend to these perturbation boundary value problems, results proved by Hale and Cesari (see, e.g., [7, 8], and Mawhin [44, 45]) for periodic solutions only, smooth solutions, and $f$ or $g$ independent of derivatives of maximum order. The following examples concerning periodic solutions may be of interest.

For periodic solutions, as well as for other particular situations, the bifurcation equations and their Jacobian matrix at $\varepsilon = 0$ have been written explicitly. The following examples are of some interest (for details see [56]):

1.

$$x'' + \omega^2 x = \varepsilon(1 - x^2)x' + \varepsilon a\omega^{-1}x'' + \varepsilon b\omega(\cos \omega t + \alpha)$$

for $\varepsilon > 0$ a small parameter, $\omega$, $a$, $b$, $\alpha$ constants, $\omega = 2\pi/T$, has periodic solutions of the form

$$x(t, \varepsilon) = \lambda(\varepsilon)\omega^{-1} \sin(\omega t + \theta(\varepsilon)) + O(\varepsilon),$$

$$\lambda(\varepsilon) = \lambda_0 + O(\varepsilon), \quad \theta(\varepsilon) = \theta_0 + O(\varepsilon),$$

where $\lambda_0$, $\theta_0$ satisfy the equations $\lambda^3 - 4p\omega^3 \cos(\alpha - \theta) = 0$, $a\lambda + p\omega \sin(\alpha - \theta) = 0$. In this example $p = q = 2$.

2.

$$x'' = \varepsilon[-e^{x''} + x^2 \sin^2 t]$$

for $\varepsilon > 0$ a small parameter has $2\pi$-periodic solutions of the form

$$x(t) = \alpha + O(\varepsilon) \quad \text{with} \quad \alpha = \pm 2^{1/2}.$$

In this example $p = q = 1$.

Hale's concept ([25], p. 267) of symmetric systems is extended in [56] to boundary value problems (7) of the form

$$\begin{aligned} x'(t) - \bar{A}(t)x &= f(t, x), \qquad -a \leqslant t \leqslant a, \\ B_1 x(-a) + B_2 x(a) &= 0. \end{aligned} \tag{10}$$

We say that (10) has a symmetry property $S$, if there is an $n \times n$ constant matrix $S$ such that (a) $S^2 = I$; (b) $Sf(-t, Sx) = -f(t, x)$; (c) if $z(t)$, $-a \leqslant t \leqslant a$, satisfies $B_1 z(-a) + B_2 z(a) = 0$, so does $Sz(-t)$; (d) if $z(t)$, $-a \leqslant t \leqslant a$, satisfies the adjoint boundary condition $\tilde{z}(-a) = \tilde{\alpha} B_1$, $\tilde{z}(a) = -\tilde{\alpha} B_2$, so does $Sz(-t)$.

Under these hypotheses, the $q$ components of the bifurcation equation can be proved to be linearly dependent [56], and thus we may expect a family of solutions to problem (10) depending on a suitable number of parameters.

The following two examples from [56] containing a small parameter $\varepsilon$ both possess a family of solutions satisfying the given boundary conditions:

1.

$$\begin{aligned} x_1' &= x_2 + \varepsilon f_1(t, x), & f_1(-t, x_1, -x_2, x_3) &= -f_1(t, x), \\ x_2' &= -x_1 + \varepsilon f_2(t, x), & f_2(-t, x_1, -x_2, x_3) &= f_2(t, x), \\ x_3' &= \varepsilon f_3(t, x), & f_3(-t, x_1, -x_2, x_3) &= -f_2(t, x), \end{aligned}$$

$$x_1(-\pi) - x_1(\pi) = 0, \quad x_2(-\pi) + x_2(\pi) = 0,$$

$$x_3(-\pi) - x_3(\pi) = 0,$$

$$S = (1, 0, 0; 0, -1, 0; 0, 0, 1), \quad x = (x_1, x_2, x_3).$$

The system has a two-parameter family of solutions $x_1$, $x_3$ even, $x_2$ odd, of the form

$$x_1 = \lambda \cos t + O(\varepsilon), \quad x_2 = -\lambda \sin t + O(\varepsilon), \quad x_3 = c + O(\varepsilon)$$

$\lambda$, $c$ arbitrary, $|\varepsilon|$ sufficiently small.

2.

$$x_1' = x_2 + \varepsilon f_1(t, x), \quad x_2' = x_3 + \varepsilon f_2(t, x), \quad x_3' = \varepsilon f_3(t, x),$$

where $-a \leqslant t \leqslant a$, $x_1(-a) - x_1(a) = 0$, $x_2(-a) + x_2(a) = 0$, $x_3(-a) - x_3(a) = 0$, $f_1$, $f_2$, $f_3$ and $S$ as in example 1. The system has a two-parameter family of solutions, $x_1$, $x_3$ even, $x_2$ odd, of the form

$$x_1 = \alpha_1 + \alpha_2 t^2 + O(\varepsilon), \quad x_2 = 2\alpha_2 t + O(\varepsilon), \quad x_3 = 2\alpha_2 + O(\varepsilon),$$

$\alpha_1$, $\alpha_2$ arbitrary, $|\varepsilon|$ sufficiently small.

## 6. PROBLEMS OF PERTURBATION TYPE FOR PARTIAL DIFFERENTIAL EQUATIONS

Shaw [65] has worked on many facets (see Section 11 below) of the problem:

$$\begin{aligned} Ex &= F(t, Dx(t)) && \text{on } G, \\ Bx &= 0 && \text{on } \partial G, \end{aligned} \tag{11}$$

where $E$ is a strongly uniformly elliptic linear operator of order $2m$ with linear homogeneous boundary conditions $Bx = 0$ on $\partial G$, under the sole and very mild hypothesis that $E$ has smooth coefficients and that the pair $(E, B)$ be coercive in the sense of Agmon, Douglis, and Nirenberg. This is an algebraic concept and most problems of applications are included. Essentially what is needed is that a Gårding inequality holds. In any case, $(E, B)$ is not necessarily selfadjoint.

Here $Dx$ denotes the set of $M$ derivatives $D^\alpha x$, $0 \leqslant |\alpha| \leqslant 2m$

and $F(t, v)$, $v = (v_1, \ldots, v_M) \in R^M$, is a continuous function defined on $\bar{G} \times R^M$. Again, $G$ is a domain in the $t$-space $R^n$ with smooth boundary $\partial G$, $t = (t_1, \ldots, t_n)$, and $\bar{G}$ is the closure of $G$.

If $(p, q)$ are the Fredholm indices of $(E, B)$, then $p$, $q$ are finite, and there are finitely many elements $\theta_i \in L_2(G)$, $i = 1, \ldots, p$, $\psi_i \in L_2(G)$, $i = 1, \ldots, q$, such that $\ker(E, B) = \mathrm{sp}(\theta_1, \ldots, \theta_p)$, $\mathrm{coker}(E, B) = \mathrm{sp}(\psi_1, \ldots, \psi_q)$. We can think of each of the two systems as being orthogonal elements in $L_2(G)$, and actually smooth of class $C^\infty(\bar{G})$. We know that $Ex = g$, $g \in L_2$, $Bx = 0$ has a solution $\bar{x}$ iff $(g, \psi_i) = 0$, $i = 1, \ldots, q$, with $(\ ,\ )$ the inner product in $L_2(G)$. For any $x \in \mathfrak{D}(E, B)$, the domain of $(E, B)$, we may write $x = v + Px$, where $v \perp \ker(E, B)$ (in the sense of $L_2$), thus, $Px = \Sigma_{i=1}^{p} c_i\theta_i$, $c_i = (x, \theta_i)$, briefly $Px = c\theta$. Analogously we define $Q$.

Let us consider the Sobolev spaces $X = H^{r+2m}(G)$ and $Y = H^r(G)$ for $r \geqslant 0$, $r > n/2$, $r$ integer, with usual norms $\|\ \|^{r+2m}$, $\|\cdot\|^r$. Let $D$ be the subspace of $X$ of all $x$ satisfying $Bx = 0$. Then, $E : D \to Y$ is defined on $D$, and a consequence of the coercivity assumption is that, for every $x \in D$ we have $\|x\|^{2m+r} \leqslant C(\|Ex\|^r + \|x\|^\circ)$, where $H^\circ$ is the $L_2$ space on $G$, $\|\ \|^\circ$ is the $L_2$-norm, and $C$ is a suitable constant.

For any $g \in \mathrm{range}(E, B)$, $g \in H^r(G)$, there are solutions $x \in H^{2m+r}$ to the problem $Ex = g$, $Bx = 0$, and if $\bar{x}$ is any such solution, all others are given by $\bar{x} + \Sigma_{i=1}^{p} c_i\theta_i$, $c_i$ arbitrary. It is proved in [65] that for one and only one $x_0$ of these solutions we have $Px_0 = 0$. The transformation $g \to x_0$ defines the operator $H$ from range $(E, B)$ into $X_0 = (I - P)H^{2m+r}$. A consequence of the assumption $r > n/2$ is that, for $x \in H^{2m+r}$, by Sobolev's imbedding theorems, $x$ and all derivatives $D^\alpha x$, $0 \leqslant |\alpha| \leqslant 2m$, appearing in the function $F$ are continuous in $\bar{G}$, and moreover their Sup norms in $\bar{G}$ are not larger than $C_\alpha\|x\|^{r+2m}$, $0 \leqslant |\alpha| \leqslant 2m$, for suitable constants $C_\alpha$. Thus, the arguments in $F$ remain bounded in the Sup norm if $\|x\|^{r+2m}$ is bounded. Since $F$ is assumed to be smooth, then $F$ is Lipschitzian in its own arguments in each fixed ball. Then $F(\cdot, Dx(\cdot))$ is a Lipschitz function of $t$ in the topology of $H^r(G)$ for $\|x\|^{r+2m}$ below any given constant $M$.

By replacing $F$ in (11) by $\varepsilon F$, where $\varepsilon$ is a small parameter, the auxiliary equation is solvable by contraction maps and Banach fixed point theorem, at least for $|\varepsilon|$ sufficiently small. It turns out that the present considerations, based on Sobolev's imbedding theorems, allow remarkable good estimates of $\varepsilon$. For $p = q = 1$, the bifurcation equation is a real equation in one real unknown. To exhibit the power of the present approach, we present here from [65] a few examples, with $p = q = 1$, $\varepsilon = 1$, one with nonlinearity including derivatives of maximal order.

1.

$$\Delta x + 2x = k - \text{arc tan } x \text{ on } G = \left[(\xi, \eta), 0 \leqslant \xi, \eta \leqslant \pi\right],$$
$$x = 0 \text{ on } \partial G.$$

Here $\lambda = 2$ is an eigenvalue of the operator $\Delta$ on $G$ with Dirichlet boundary conditions, and eigenfunction $\sin \xi \sin \eta$, $(\xi, \eta) \in G$. It was shown in [65] that the problem is solvable if and only if $|k| < \pi/2$. Moreover, for $|k| < \pi/2$, the solution is then of the form $x(\xi, \eta) = v_d(\xi, \eta) + d \sin \xi \sin \eta$, with $\int_G v_d(\xi, \eta) \sin \xi \sin \eta \, d\xi d\eta = 0$, $\|v_d\|^\circ \leqslant \pi^2/3$, and a suitable value of the constant $d$.

2.

$$\Delta x = \sin 2\pi\xi + 2^{-1/2} x^{1/3}\left(x^2(1 + x^2)^{-1}\right) + \cos x_{\xi\eta}, \; (\xi, \eta) \in G,$$
$$\partial x / \partial n = 0 \text{ on } \partial G, \quad G = \left[(\xi, \eta) | 0 \leqslant \xi, \eta \leqslant 1\right].$$

The problem has a solution of the form $x(\xi, \eta) = d + v(\xi, \eta)$, $d$ a suitable constant, and $\int_G v(\xi, \eta) \, d\xi d\eta = 0$.

3.

$$\Delta x = f(\xi, \eta) \pm 10^{-1} x^3 \text{ on } G = \left[(\xi, \eta) | 0 \leqslant \xi, \eta \leqslant 1\right],$$
$$\partial x / \partial n = 0 \text{ on } \partial G,$$

where $f$ is a given measurable bounded function on $G$. The problem has a solution for $|f| \leqslant 1/20$.

## 7. REDUCTION TO AN ALTERNATIVE PROBLEM BY MONOTONE MAPS

From a completely different angle, Cesari and Kannan [11] have considered the equation $Ex = Nx$ in the case where $E : \mathfrak{D}(E) \to X$, $\mathfrak{D}(E) \subset X$, and $N : X \to X$ are operators in a Hilbert space $X$, $\mathfrak{D}(E)$ everywhere dense in $X$, and $N : X \to X$ is monotone. Under mild hypotheses they proved that the auxiliary equation is uniquely solvable, and $Ex = Nx$ is reducible to an alternative problem by Brézis' and Minty's monotone and maximal monotone operator theory [1, 53]. A further relevant point is that under mild assumptions and the use of Brézis, Crandall, and Pazy [2] and Kato's theorems [33], even the bifurcation equation in a finite dimensional space (alternative problem) can be solved, and therefore the original problem $Ex = Nx$ is solvable.

We just mention here that an operator $N$, possibly nonlinear, in a real Hilbert space $X$, that is, $N : \mathfrak{D}(N) \to X$, $\mathfrak{D}(N) \subset X$, is said to be monotone if $x, y \in \mathfrak{D}(N)$ implies $(x - y, Nx - Ny) \geqslant 0$, where $( \, , \, )$ is the inner product in $X$. More generally, a set-valued operator $N : \mathfrak{D}(N) \to 2^X$, mapping each element $x$ of its domain $\mathfrak{D}(N) \subset X$ into a subset $N(x)$ of $X$, is said to be monotone if $x, y \in \mathfrak{D}(N)$, $u \in N(x)$, $v \in N(y)$ implies $(x - y, u - v) \geqslant 0$. Thus, the property of the operator $N : \mathfrak{D}(N) \to 2^X$ to be monotone is actually a property of the graph $G(N)$ of $N$, that is, of the set $G(N)$ of all pairs $(x, u) \in X \times X$, $x \in \mathfrak{D}(N)$, $u \in N(x)$. Thus, an operator $N : \mathfrak{D}(A) \to 2^X$, is said to be maximal monotone if $N$ is monotone and its graph $G(N)$ is not the proper subset of some larger subset of $X \times X$ which is monotone. Moreover, a map $N : \mathfrak{D}(N) \to 2^X$, from its domain $\mathfrak{D}(N) \subset X$ onto its range $\mathfrak{R}(N) \subset X$ is monotone [maximal monotone] if and only if its inverse $N^{-1} : \mathfrak{R}(N) \to 2^X$ from $\mathfrak{R}(N)$ onto $\mathfrak{D}(N)$ is monotone [maximal monotone]. An operator $N : \mathfrak{D}(N) \to X$ defined on a convex domain $\mathfrak{D}(N)$ of $X$ is said to be hemicontinuous if $N$ is continuous from every segment $s$ in $\mathfrak{D}(A)$ to $X$, the strong topology on $s$ and the weak topology on the range.

Of the various statements, we report here the following ones which include a resonance case in abstract form for selfadjoint problems:

THEOREM 1 (Cesari and Kannan [11]): *Let $E$ be selfadjoint with eigenvalues $\lambda_i$ and eigenfunctions $\phi_i$ $i = 1, 2, \ldots, E\phi_i + \lambda_i\phi_i = 0, \lambda_i \to +\infty$ as $i \to +\infty$. If $N : X \to X$ is monotone and hemicontinuous, then the auxiliary equation has always one and only one solution $x = \mathfrak{T}(x^*)$, $Px = x^*$ for every $x^* \in X_0 = \ker E$. Thus, problem (1) is reducible to an alternative one.*

THEOREM 2 (Cesari and Kannan [11]): *Under the assumptions above, if $\lambda_i \geqslant 0$, $i = 1, 2, \ldots,$ are the eigenvalues of $E$, $E\phi_i + \lambda_i\phi_i = 0$, $\lambda_i \to +\infty$ as $i \to +\infty$, and $N : X \to X$ is monotone, hemicontinuous, and coercive, then the alternative problem has also a unique solution, and so has the given problem $Ex = Nx$. The coercivity condition can be removed if we know, instead, that there is some $R > 0$ such that $\langle Nx, x^* \rangle \geqslant 0$ for all $x \in X$, $x^* \in X_0$, $Px = x^*$, $\|x^*\| = R$.*

Theorem 1 was proved in [7] showing that the relevant map is the sum of two maximal monotone maps and also maximal monotone by a theorem of Rockafellar (see Brézis, Crandall, Pazy [2]). Theorem 2 under the main hypothesis was proved in [11], and in more details in [7, 9].

The theorems above concerning the general reduction of problems to the bifurcation equation, and the solvability of the auxiliary and bifurcation equation have been extended to the "nonselfadjoint" case by Gustafson and Sather [21, 22] and by Osborn and Sather [60]. For what concerns the use of monotone operators, Theorem 1 above has been extended to nonselfadjoint problems by Osborn and Sather [61] and Cesari [8], and Theorem 2 (main assumption case) by Nagle [56].

The requirement $N : X \to X$, that is, that $N$ is defined on the whole of $X$ is often not satisfied. One can think of $X = L_2(G)$, $G$ some domain in $E^\nu$, $\nu \geqslant 1$, and $Nx$ the Nemitski operator $Nx(t) = F(t, x(t))$, and $F$ a polynomial in $x$. To overcome this difficulty much work has been done by Gustafson and Sather [21, 22] by the use of the square root decomposition $K = K^{\frac{1}{2}}K^{\frac{1}{2}}$ of a positive compact operator $K$. In the same line Kannan and Locker [31] have shown that any known decomposition of the relevant operators, say $E = TT^*$, $K = J^*J$, into operators $T$, $T^*$ or $J^*$, $J$, whose boundary conditions match those we have already on $E$, or $K$, can be used to the same purpose. We refer to their paper [31] for a great many details and applications.

In other developments Kannan [28, 29, 30] has used Schauder's principle of invariance of domain to derive further existence theorems at resonance. We refer to Nagumo [57, 58] for Schauder's principle.

## 8. CONNECTIONS WITH THE LERAY-SCHAUDER APPROACH AND WORK OF WILLIAMS AND OF MAWHIN

First, it is useful to write the auxiliary and the bifurcation equations as:

$$y = T_1(x, x^*) = x^* + H(I - Q)Nx,$$

$$y^* = T_2(x, x^*) = x^* + Q(E - N)x,$$

$$T = (T_1, T_2),$$

that is, a transformation $(x, x^*) \to (y, y^*)$ in the product space $X \times X_0$, or total map, for which now we search for fixed points $y = x$, $y^* = x^*$. Under mild hypotheses and actually in all applications we shall mention below, this transformation has compactness properties, so that Schauder's fixed point theorem may be applicable. Also a Leray–Schauder degree $d_{LS}[I - T, \Omega, 0]$ in some large ball $\Omega$ of the domain can be defined.

Landesman and Lazer [38] used the total map $T$ and Schauder's fixed point theorem in the original proof of their theorem, as did Williams [67] shortly after, by a simpler argument. Cesari has been using the same total map $T$ and Williams argument via Schauder's fixed point theorem in his proof of the recent De Figueiredo extension of the Landesman and Lazer theorem (see Section 10). In the study of the forced oscillations of Liénard systems, Cesari and Kannan ([12, 13]) have used again the total map $T = (T_1, T_2)$ and its compactness properties (see Section 7), have given *a priori* bounds for the solutions, and made use of Borsuk–Ulam lemma (dispensing therefore with the explicit use of the topological degree —which of course is $\neq 0$ for the simple reason that a certain relevant map is odd).

In the selfadjoint case ($X = Y, P = Q$) Williams [66] considered the map $W$ defined by

$$Wz = Pz + H(I - P)Nz - P(E - N)z,$$

and noted that the fixed points of $W$ are exactly the solutions of $Ex = Nx$. Williams worked in situations where $T_1 = P + H(I - P)N$ is a contraction on the fibers $P^{-1}x^*$ of $P$, but $T_1$ is not necessarily compact. Williams noted that, in such a situation, the composite map $W'$ defined by $W'z = W\mathfrak{T}z$ has the same fixed points as $W$, that is, exactly the solutions of $Ex = Nx$, but now $W'$ is compact, and a Leray–Schauder degree $d_{LS}[I - W', \Omega, 0]$ can be defined. Williams proved the following remarkable theorem:

THEOREM (Williams [66], 1968): *For $T_1$, a contraction on the fibers $P^{-1}x^*$ of $P$, the Leray–Schauder degree just mentioned is equal to the Brouwer degree of the mapping $T_2\mathfrak{T}$, or briefly $T_2$, of the sole bifurcation equation in a finite dimensional domain $\Omega_0$, and the equality holds*

$$d_{LS}[I - W', \Omega, 0] = d_{Br}[T_2, \Omega_0, 0].$$

Later, Mawhin ([49], 1972, and subsequent work) has reconsidered the map $W$ for nonselfadjoint problems, with $X_0 = \ker E$, $Y = \operatorname{coker} E$. If $S : Y_0 \to X_0$ is any continuous map, linear or nonlinear, with the property that $S^{-1}(0) = 0$, then for the map $W$ defined by

$$Wz = Pz + H(I - Q)Nz - SQN,$$

the fixed points of $W$ are exactly the solutions of the original problem $Ex = Nx$. In the problems taken into consideration by Mawhin, $W$ is always a compact map because the total map itself is already a compact map in the product space.

Then a Leray–Schauder degree $d_{LS}(I - W, \Omega, 0)$ can be defined, and it can be estimated under suitably *a priori* bounds. This is Mawhin's coincidence degree.

One of Mawhin's applications is the study of aspects of forced oscillations of Liénard systems. As mentioned above (see also Section 10 below), Cesari and Kannan ([12, 13]) have made a further study of these oscillations, by using the total map $T$ and the Borsuk–Ulam lemma in lieu of topological degree. Another application of Mawhin's coincidence degree is a study of the Landesman and Lazer theorem. As mentioned above (see also Section 11 below), Cesari has used directly (see [9]) the total map for a proof in the line of Williams, of recent De Figueiredo results, which are a far-reaching extension of the Landesman and Lazer theorem.

Work is going on at present to establish precise relations between the degree of the total map in the product space, the degree of $W$, that is, the coincidence degree, and of course the Brouwer degree of the relevant finite dimensional maps whenever they are defined. This also in connection with recent extensions due to Rothe of the concept of degree and its properties, and parallel work of Zabreiko and Strygina in connection w[illegible] the alternative method.

The previous analysis seems to indicate that Mawhin's coincidence degree argument is well within the frame and structure of the alternative method and the use of the Brouwer and Leray–Schauder topological degrees.

## 9. STOCHASTIC PROBLEMS

In stochastic operator equations one often obtains the existence of a random solution as follows: For each $\omega$ in the probability space one obtains the existence of a solution by deterministic methods, namely by iterative techniques, and then one establishes the randomness with the help of the iterative process. However, one encounters problems in random operator equations where the existence of a solution for each $\omega$ is obtained by means of, say, the Schauder fixed point theorem, or the Leray–Schauder degree theory, and the measurability property cannot be proved directly. Kannan and Salehi [32] have overcome this difficulty by showing by different methods that solutions so obtained are measurable in $\omega$, and thus random solutions according to the usual terminology. Let us consider the usual equation $Ex = Nx$, where $E$ is a selfadjoint linear differential operator, and $N$ is now a random nonlinear operator over the real Hilbert space $X$. Let $\lambda_i \geqslant 0$, $\lambda_i \leqslant \lambda_{i+1}$, $\lambda_i \to \infty$, be the eigenvalues of $E$. When $\lambda = 0$ is an eigenvalue, and $N$ satisfies suitable monotonicity conditions, then as we have seen before (see Section 7 above), for each $\omega$ the equation over $(\ker E)^{\perp}$ (the auxiliary equation) is solved by Minty's [53] and Brézis's [1] theory of monotone operators, and the equation over $\ker E$ (the bifurcation equation) may be solved by Leray–Schauder degree theory, yielding an equation $f(\omega, x) = 0$. The existence of a random solution is then proved by the study of the graph of $f$, and from results of the theory of randomness of inverse operators.

Here is one of the existence theorems by Kannan and Salehi [32]. Here $\lambda_{m+1}$ denote the smallest eigenvalue of $E$ which is $> 0$.

THEOREM: *If* $N : \Omega \times X \to S$ *is continuous and bounded, if there is a constant* $\mu$, $0 \leqslant \mu < \lambda_{m+1}$, *such that* $(Nx - Ny, x - y) \geqslant -\mu\|x - y\|^2$ *for all* $x, y \in X$; *and if finally there is a real* $R > 0$ *such that for all* $x \in X$, $x^* \in X_0$ *with* $\|x^*\| = R$, $Px = x^*$, *we have* $(Nx, x^*) \geqslant 0$, *then problem* $Ex = Nx$ *has a random solution.*

## 10. SOLUTIONS IN THE LARGE OF LIÉNARD SYSTEMS WITH FORCING TERMS

We consider here nonlinear Liénard systems with periodic forcing terms of the type

$$x'' + (d/dt)\text{grad } G(x(t)) + (d/dt)V(x(t), t) + Ax(t) + g(x(t)) = e(t), \qquad (12)$$

where $x = (x_1, \ldots, x_n)$, $V(x, t) = (V_1, \ldots, V_n)$, $e(t) =$

$(e_1, \ldots, e_n)$, $g(x) = (g_1, \ldots, g_n)$, $G(x)$ a scalar, $A = [a_{ij}]$ a constant $n \times n$ matrix, and where we assume $e(t)$ to be $2\pi$-periodic and of mean value zero, and $g(x)/|x| \to 0$ as $|x| \to \infty$. We also assume $G$ of class $C^2$, $V$ of class $C^1$, $g$ continuous, $e$ continuous, $V$ $2\pi$-periodic in $t$. Here is a statement we have recently proved:

THEOREM 1 (Cesari and Kannan [13]): *Under the assumptions above, system* (11) *has at least a $2\pi$-periodic solution $x(t) = (x_1, \ldots, x_n)$ provided one of the following assumptions holds: either*

(A) *$G(x)$ is homogeneous in $x$ of degree $2p$, of constant sign, satisfying $|G(x)| \geqslant c|x|^{2p}$ for some integer $p \geqslant 2$, constant $c > 0$, and all $x \in R^n$; $|V(x, t)| \leqslant C|x|^p + D$ for some constants $C$, $D \geqslant 0$ and all $(x, t) \in R^{n+1}$; $A$ is nonsingular; $g(x)/|x| \to 0$ as $|x| \to \infty$;*

*or* (B) *the same as* (A) *with $p = 1$ and $2c > C + \|A - A_{-1}\|/2$, $A$ nonsingular, $g(x)/|x| \to 0$ as $|x| \to \infty$;*

*or* (C) *the same as* (A) *with $p \geqslant 1$, with $V(x)$ depending on $x$ only of the form $V(x) = \text{grad } W(x)$, $W$ of class $C^2$ in $R^n$, $|\text{grad } W(x)| \leqslant C|x|^{2p-2} + D$ for some constants $C$, $D \geqslant 0$, and all $x \in R^n$; $A$ nonsingular, $g(x)/|x| \to \infty$ as $|x| \to \infty$, and $2c > \|A - A_{-1}\|/2$ if $p = 1$;*

*or* (D) *$V \equiv 0$, $A \equiv 0$, $n = 1$, $G : R \to R$ an arbitrary function of class $C^2$ on $R$, $g(x)/|x| \to 0$ as $|x| \to \infty$, $xg(x) \geqslant 0$ for all $|x| \geqslant a$ and some constant $a > 0$;*

*or* (E) *$V \equiv 0$, $A \equiv 0$, $n > 1$, $G : R^n \to R^n$ an arbitrary function of class $C^2$ in $R^n$, $g(x)/|x| \to 0$ as $|x| \to \infty$, $g(x) \neq 0$ for all $|x| \geqslant a$, and the geometrical angle between $x$ and $g(x)$ in $R^n$ is $\leqslant \pi/2 - \sigma$, for some constants $a > 0$, $0 < \sigma < \pi/2$, and all $x \in R^n$, $|x| \geqslant a$.*

The case $g \equiv 0$ gives rise to alternate and slightly different conditions:

THEOREM 2 (Cesari and Kannan [12, 13]): *Under the same general assumptions of Theorem* 1, *the same conclusion holds provided one of the following assumptions holds: either*

(a) *$g \equiv 0$, $G$ and $V$ as in* (A) *with $p \geqslant 2$, $A$ an arbitrary constant matrix;*

*or* (b) $g \equiv 0$, $G$ *and* $V$ *as in* (A) *with* $p = 1$ *and* $2c > C + \|A - A_{-1}\|/2$, $A$ *an arbitrary constant matrix*;

*or* (c) $g \equiv 0$, $G$ *and* $V$ *as in* (C) *with* $p \geqslant 1$, $A$ *an arbitrary constant matrix*, *and* $2c > \|A - A_{-1}\|/2$ *if* $p = 1$;

*or* (d) $g \equiv 0$, $V \equiv 0$, $G : R^n \to R$ *an arbitrary function of class* $C^2$; $A$ *definite negative*;

*or* (e) $g \equiv 0$, $V \equiv 0$, $G : R^n \to R$ *an arbitrary function of class* $C^2$, $A$ *an arbitrary constant matrix with* $\|A\| < 1$.

In 1951 Graffi [20] proved the existence of periodic solutions of Liénard systems with forcing terms, on the basis of Brouwer's fixed point theorem, thus extending previous work of Lefschetz [39], Levinson [41], and Mizohata and Yamaguti [55]. The main point of their argument is the geometrical property that, under hypotheses, all solutions from some closed ball in the phase space $E^{2n}$ will return to it within one period. Graffi's geometric approach was continued by Mizohata [54]. Graffi also proved [19] *a priori* bounds in the square norm for the periodic solutions. Independently, and much more recently, Lazer [37] proved the existence of periodic solutions for the second order equation $x'' + cx' + g(x) = e(t)$, $e$ periodic. Graffi's *a priori* bounds were then used by Mawhin [47, 48] to prove existence theorems in the frame of the discussed above alternative method, and thereby improving on Lazer's results and on some of the work of Graffi. In Cesari's and Kannan's papers [12, 13], the results of Mawhin's are further extended, and Theorems 1 and 2 give a sample of the results.

Concerning Theorem 1, the proof, as mentioned, is based on the study of the total map $T = (T_1, T_2)$ in the product space $(L_2([0, 2\pi]))^n \times R^k$, $k$ the dimension of ker $E$, a space of constants, in the scheme corresponding to selfadjoint problems, $X = Y$, $P = Q$. The use of the Kannan–Locker decomposition $-H(I - P) = J^*J$ helps in proving that $T$ is compact. Then, difficult estimates give rise to $L_2$-bounds on the derivatives of the solutions, and of course Sup bounds on the solutions themselves. In other words, bounds are obtained for the solutions $x$ in the Sobolev space $H^1$ of periodic functions. These estimates are an improvement on those of Graffi [19] and [47, 48]. At this point arguments by topological degree, or by coincidence degree, would be quite natural. Cesari and Kannan have preferred to conclude the proof of Theorem 1 by the use of the Borsuk–Ulam lemma. We state here two forms of this lemma: the second form was actually used in [12, 13].

(a) *Let $X$ be a Banach space, $F : X \to X$ a compact map, $F$ odd outside some ball, that is, there is $R \geqslant 0$ such that $F(-x) = -F(x)$ for all $x \in X$, $\|x\| \geqslant R$, and let us assume that $(I + F)^{-1}$ maps bounded sets of $X$ into bounded sets of $X$. Then $(I + F)X = X$.*

(b) (An extended form of (a)). *Let $X$ be a Banach space, let $F_0, F_1 : X \to X$ be compact maps, $F_0$ odd outside some ball in $X$, let $F = F_0 + F_1$, and let us assume that there is some function $k(\xi) \geqslant 0$, $0 \leqslant \xi < \infty$, such that whenever $x + (F_0 + \lambda F_1)x = w$ for some $x, w \in X$, $0 \leqslant \lambda \leqslant 1$, then we also have $\|x\| \leqslant k(\|w\|)$. Then $(I + F)X = X$.*

Concerning Theorem 2 on Liénard systems, we note here that, under its assumptions, in particular $g \equiv 0$, the solutions $x$ of the problem have mean value zero. In other words, under the natural choice of $P$, the bifurcation equation is trivially satisfied ($x^* = Px = 0$), and we are reduced to the only auxiliary equation. Cesari and Kannan [12, 13] proved the expected compactness properties of $T_1$, proved suitable *a priori* bounds, and again applied the Borsuk–Ulam lemma (extended form (b)).

## 11. RECENT DEVELOPMENTS IN THE LANDESMAN AND LAZER THEOREM

**(a) The selfadjoint case.** Let us consider the nonlinear problem

$$Ex = F(t, Dx(t)), \quad t \in G,$$
$$x = 0, \quad \partial x/\partial \nu = 0, \ldots, \quad \partial^{m-1}x/\partial \nu^{m-1} = 0 \quad \text{on } \partial G, \quad (13)$$

where $Dx$ denotes the set of all derivatives $D^\alpha x = \partial^{|\alpha|}x/\partial t_1^{\alpha_1} \cdots \partial t_n^{\alpha_n}$, $\alpha = (\alpha_1, \ldots, \alpha_n)$, $0 \leqslant |\alpha| \leqslant 2m - 1$, and where $(E_1)$ $E$ is a linear uniformly elliptic selfadjoint operator of order $2m$ with smooth coefficients on a domain $G$ of the $t$-space $E^n$, $t = (t_1, \ldots, t_n)$. Here $G$ has a smooth boundary $\partial G$, and $\nu$ is the inner normal on $\partial G$. Let us assume that the remaining few assumptions are satisfied. First we explicitly assume that $(E_2)$ the null space $W$, or $W = \ker E$, is nontrivial, of dimension $k$, $1 \leqslant k < \infty$, and we denote by $w = (w_1, \ldots, w_k)$, an arbitrary base for $W$. For $c = (c_1, \ldots, c_k) \in R^k$, then $v(x) = cw = c_1w_1 + \cdots + c_kw_k$ spans $W$.

Let $M$ denote the total number of distinct derivatives $D^\alpha u$, $0 \leqslant |\alpha| \leqslant 2m - 1$, or $M = 1 + n + n(n+1)/2 + \cdots$, let $\tilde{v}$ denote the $M$-vector $\tilde{v} = (v_1, \ldots, v_M) \in R^M$, and let $F(t, \tilde{v})$ denote a given function on $\bar{G} \times R^M$, $\bar{G}$ the closure of $G^M$. We assume that $F$ satisfies a Carathéodory condition, that is, $F$ is measurable in $t$ for every $\tilde{v}$, and continuous in $\tilde{v}$ for every $t$. Let $\Sigma$ denote the unit sphere in $R^M$, or $\Sigma = [\eta \in R^M, \|\eta\| = 1]$.

We just mention here that, in terms of Sobolev spaces, weak solutions of the linear underlying problem $Ex = g(t)$, $t \in G$, $g \in L_2(G)$, corresponding to the boundary conditions stated above, are simply required to belong to the Sobolev space $H_0^m(G)$. Also, the problem $Ex = g(t)$, $g \in L_2(G)$, has a weak solution $x \in H_0^m$ if and only if $g$ is orthogonal to the $k$ functions $w_i$, $i = 1, \ldots, k$. Furthermore, because of the assumed smoothness of $\partial G$ and of the coefficients of the uniformly elliptic linear operator $E$, any such weak solution $x \in H^m$ is actually of Sobolev class $H^{2m}$, and thus $x \in H_0^m \cap H^{2m}$ (Stampacchia's and Morrey's regularity property).

For $m = 2$, $k = 1$, and $F$ of the form $F = f(x) + h(t)$, $x \in E^1$, $t \in G$, $h \in L_2(G)$, problem (13) reduces to

$$Ex(t) = f(x(t)) + h(t) \quad \text{on } G,$$
$$x(t) = 0 \quad \text{on } \partial G. \tag{14}$$

Let us assume that $f$ possesses finite limits $R = f(+\infty)$, $r = f(-\infty)$, and that $r \leqslant f(s) \leqslant R$ for all $s$ real. Let us denote by

$w(t)$, $t \in G$, any eigenfunction of the underlying linear problem, by $G^+$, $G^-$ be the sets of all points $t \in G$ where $w(t) \geqslant 0$, $w(t) \leqslant 0$, respectively, and by $w^+$, $w^-$ the nonnegative numbers

$$w^+ = \int_{G^+} |w|\, dt,\ w^- = \int_{G^-} |w|\, dt.$$

Then Landesman and Lazer [38] proved that

$$rw^+ - Rw^- \leqslant (h, w) \leqslant Rw^+ - rw^-$$

is a necessary condition for (14) to have a weak solution $x \in H_0^1(G)$. They also proved, by the alternative method, that the slightly stronger condition

$$rw^+ - Rw^- < (h, w) < Rw^+ - rw^- \tag{15}$$

is sufficient for (14) to have a weak solution $x \in H_0^1(G)$.

Williams [67] considered again problem (14) with $E$ an elliptic operator as above but with $m \geqslant 1$, $k \geqslant 1$. Again by the alternative method Williams proved that, if (15) holds for any eigenfunction $w$, then (14) has at least one weak solution $x \in H_0^m$.

Let us now return to the general problem (13) with $E$ an elliptic operator as above, this time under smoothness hypotheses on $\partial G$ and on the coefficients, with $m \geqslant 1$, $k \geqslant 1$, and $F(t, \tilde{v})$ any function in $\bar{G} \times E^M$, satisfying a Carathéodory condition. Let us assume, with De Figueiredo, that the following assumptions are satisfied:

$(F_1)$ For every $t \in \bar{G}$, $\eta \in \Sigma$, and sequences $r_n$, $\eta_n$ with $r_n > 0$, $r_n \to \infty$, $\eta_n \in \Sigma$, $\eta_n \to \eta$ as $n \to \infty$, we have

$$\lim_{n\to\infty} F(t, r_n\eta_n) = h(t, \eta);$$

$(F_2)$ The function $h(t, v)$, $(t, v) \in \bar{G} \times R^M$, satisfies a Carathéodory condition, and there are functions $g_0(t)$, $g(t)$, $t \in G$, $g_0$, $g \in L_2(G)$, such that

$$|h(t, \eta)| \leqslant g(t) \quad \text{for all } t \in G, \eta \in \Sigma,$$

$$|F(t, \tilde{v})| \leqslant g_0(t) \quad \text{for all } t \in G, \tilde{v} \in R^M;$$

$(F_3)$ For every $v(t) = bw = b_1w_1 + \cdots + b_kw_k$, $|b| = 1$, we have

$$\int_G h\left(t, \frac{Dv(t)}{|Dv(t)|}\right)v(t)\,dt > 0.$$

THEOREM 1: *Let E be selfadjoint and satisfy* $(E_1)$ *and* $(E_2)$, *with null space W of dimension* $k \geq 1$. *Let* $F(x, \tilde{v})$ *be a function on* $\bar{G} \times R^M$ *satisfying a Carathéodory condition and assumptions* $(F_1)$, $(F_2)$, $(F_3)$. *Then, problem* $Ex = F(t, Dx(t))$, $x \in G$, *has at least one solution* $x \in H_0^m \cap H^{2m}$, *with* $Nx = F(\cdot, Dx(\cdot)) \in L_2(G)$.

For a slightly more general form of this theorem, let $p$ be any number, $0 \leq p < 1$, and let us denote by $F_{1p}$, $F_{2p}$, $F_{3p}$ the following conditions on $F$:

$(F_{1p})$ For every $t \in G$, $\eta \in \Sigma$, and sequences $r_n$, $\eta_n$ with $r_n > 0$, $r_n \to \infty$, $\eta_n \in \Sigma$, $\eta_n \to \eta$ as $n \to \infty$, we have

$$\lim_{n\to\infty} r_n^{-p}F(t, r_n\eta_n) = h(t, \eta);$$

$(F_{2p})$ The function $F(t, v)$, $(t, v) \in \bar{G} \times R^M$, satisfies a Carathéodory condition, and there are functions $g(t)$, $g_0(t)$, $g_1(t)$, $t \in G$, $g_0 \in L_2(G)$, $g, g_1 \in L_{2/(1-p)}(G)$, such that

$$|h(t, \eta)| \leq g(t) \quad \text{for all } t \in G, \eta \in \Sigma,$$
$$|\hat{v}|^{-p}|F(t, \tilde{v})| \leq g_0(t) + |v|^p g_1(t) \quad \text{for all } t \in G, \tilde{v} \in R^M;$$

$(F_{3p})$ For every $v(t) = bw = b_1w_1 + \cdots + b_kw_k$, $|b| = 1$, we have

$$\int_G h(t), \frac{Dv(t)}{|Dv(t)|}\, |Dv(t)|^p v(t)\,dt > 0.$$

THEOREM 2: *Let E be selfadjoint and satisfy* $(E_1)$ *and* $(E_2)$, *with null space W of dimension* $k \geq 1$. *Let* $F(t, \tilde{v})$ *be a function on*

*$\bar{G} \times R^M$ satisfying a Carathéodory condition and assumptions $(F_{1p})$, $(F_{2p})$, $(F_{3p})$ for a given $p$, $0 \leq p < 1$. Then, problem $Ex = F(t, Dx(t))$, $t \in G$, has at least one solution $x \in H_0^m \cap H^{2m}$, with $Nx = F(\cdot, Dx(\cdot)) \in L_2(G)$.*

For $p = 0$, Theorem 2 reduces to Theorem 1. For $F$ of the form $F = f(x) + g(t)$, $x \in R^1$, $t \in G$, Theorem 1 reduces essentially to Williams' result, and for $m = 1$, $k = 1$, to Landesman's and Lazer's theorem (solutions in $H_0^m \times H^{2m}$ instead of weak solutions in $H_0^m$ only because Stampacchia's and Morrey's regularity property is used).

Both Theorems 1 and 2 are very recent. They have been proved by De Figueiredo [16] by a variant of a process of successive approximations which had been first proposed by Hess [28] for another extension of the Landesman and Lazer theorem. Theorems 1 and 2 have also been proved by Cesari (see [9]) by the use of the alternative method, namely, by the same argument used by Williams [67] in his proof.

Following Williams [67] we write the auxiliary equation and the bifurcation equation in the form of a system, say

$$\bar{x} = cw + H(I - P)Nx,$$

$$\bar{c} = c - (N\bar{x}, w),$$

where $(N\bar{x}, w) = [(N\bar{x}, w_i), i = 1, \ldots, k]$ are usual inner products in $L_2(G)$. Any fixed point $(x, c)$ of this transformation, say $c \in R^k$, $x \in H_0^m \cap H^{2m}$, is a solution of the original problem $Ex = Nx$, $u \in H_0^m \cap H^{2m}$. Indeed, for $\bar{x} = x$, $\bar{c} = c$, we have

$$x = cw + H(I - P)Nx, \quad (Nx, w) = 0.$$

The second equation shows that $PNx = 0$, and by applying $E$ to the first equation we have $Ex = 0 + (I - P)Nx = Nx$.

We may well think of $T$ as a transformation of $H^{2m-1} \times R^k$ into $H^{2m-1} \times R^k$. Indeed, for $x \in H^{2m-1}$, then all components of $\tilde{v} = Dx = [D^\alpha x, 0 \leq |\alpha| \leq 2m - 1]$ are at least in $L_2(G)$, and the mappings $x \to D^\alpha x$ are bounded linear maps from $H^{2m-1}$ into $L_2(G)$. Having assumed that $|F(t, v)| \leq C|v| + D$, $C$, $D$ constants we see that $Nx = F(t, Dx(t))$, $t \in G$, is also an element of $L_2$, and $x \to Nx$ is a bounded, though in general nonlinear map. Then, $(I - P)Nx$ is in $D$, the orthogonal complement of $W$ in $L_2$. Hence, $H(I - P)Nx$ is defined, is an element of $H_0^m \cap H^{2m}$, and also an element of $W^{\perp}$. Since $cw$ also belongs to $H_0^m \cap H^{2m}$, actually in the complementary space $W$, then $x^* \in H_0^m \cap H^{2m}$. Then, again $x^* \in H^{m-1}$, $Nx^* \in L_2$, and $(Nx^*, w) = [(Nx^*, w_i), i = 1, \ldots, k]$ is well defined and $c^* \in R^k$.

Actually, $T : H^{2m-1} \times R^k \to H^{2m-1} \times R^k$ is compact. Indeed, we have just seen that $x^* \in H^{2m}$, the inclusion map $j : H^{2m} \to H^{2m-1}$ is compact, $F(t, Dx(t))$ is compact as a continuous map on a compact map, and finally $R^k$ is finite dimensional.

At this point, the use of relation $(F_3)$ shows that some convex subset of $H^{2m-1} \times R^k$ is mapped into itself by $T$, and thus $T$ has at least a fixed point by Schauder's fixed point theorem.

This, which was also the original line of proof of Landesman and Lazer, was simplified by Williams to prove his version of the theorem. The author has just showed that the same Williams' argument, without losing its simplicity, extends to the present rather general situation of Theorems 1 and 2 above.

Theorems 1 and 2 which, as stated, have been also proved by De Figueiredo, improved previous results by De Figueiredo [14, 15] himself, by Mawhin and Franchetta [50]. The original research of Landesman and Lazer originated a great deal of work. I mention here the work of Necas [59], who used the same kind of argument, and of Mawhin and Franchetta [50], and of Hess [27], Fucik [17], Fucik, Kucera, and Necas [18]. Recently, Shaw [65] proved, by the alternative method, Landesman–Lazer type theorems for nonselfadjoint nonlinear elliptic problems (see Part b).

**(b) The nonselfadjoint case. Work of Shaw.** We present now Shaw's contributions to Landesman and Lazer type results for nonselfadjoint problems.

Let us consider, as in Section 4, a boundary value problem of the form

$$\begin{aligned} Ex &= F(t, x(t)), \quad t \in G, \\ Bx &= 0 \quad \text{on } \partial G, \end{aligned} \tag{16}$$

where $E$ is a strongly uniformly elliptic linear operator of order $2m$ with linear homogeneous boundary conditions $Bx = 0$ on $\partial G$, where $G$ is a domain in the $t$-space $E^n$, $t = (t_1, \ldots, t_n)$ with smooth boundary $\partial G$, where $E$ has smooth coefficients on $\bar{G}$, and $(E, B)$ is a coercive pair in the sense of Agmon, Douglis, and Nirenberg. Here $F$ is a given continuous function on $\bar{G} \times E^1$. We refer to Section 4 for remaining notations.

First let us consider the case of Fredholm indices (1, 1). We may think of ker $E$ as generated by a nonzero function $\theta(t)$, $t \in G$, and the coker of $E$ by a nonzero function $\psi(t)$, $t \in G$. It is relevant in the analysis of Shaw that $\theta$ and $\psi$ share regions of positivity and negativity, so that $G$ is the disjoint union of sets $G_+, G_-, G_0$ where both $\theta$ and $\psi$ are $\geqslant 0$, or $\leqslant 0$, or zero, respectively. Under this hypothesis, let $F$ be of the form $F = f(x) + g(t, x)$, $t \in G$, $x \in E^1$, $f, g$ bounded, let us assume that $f$ possesses finite limits $R = f(+\infty)$, $r = f(-\infty)$, $R \geqslant r$, and let us take $\alpha = \sup|g(t, x)|$. If

$$rw^+ - Rw^- < -\alpha \int_G |w| \, dt \leqslant \alpha \int_G |w| \, dt < Rw^+ - rw^-, \tag{17}$$

then the problem $Ex = Nx$ in $G$, $Bx = 0$ on $\partial G$, has at least one solution $x \in H^{2m}(G)$ (Shaw [65]). In particular, if $F$ is of the form $F = f(x) + g(t)$, $t \in G$, $x \in E^1$, $g \in L_2(G)$, then [65] relation (16) can be replaced by

$$rw^+ - Rw^- < \int_G g\psi \, dt < Rw^+ - rw^-. \tag{18}$$

The condition that $\theta$ and $\psi$ share regions of positivity is certainly satisfied when the pair $(E, B)$ is selfadjoint. Thus, the above result by Shaw contains as particular cases those of Williams and of Landesman and Lazer.

A further analysis has shown that for, $p = q = 1$, the function $\theta$ and $\psi$ share regions of positivity in a wide range of situations, say ordinary differential operators of the form $a_0(t)x'' + a_1(t)x' + a_2(t)x$ on $[a, b]$ with $a_0(t) > 0$, $a_0, a_1, a_2$ continuous, and Dirichlet data $x(a) = x(b) = 0$. The same is true for strongly elliptic partial differential operators of the form

$$a_0(\xi)x_{\xi\xi} + b_0(\eta)x_{\xi\xi} + a_1(\xi)x_\xi + b_1(\eta)x_\eta + c(\xi, \eta)x,$$

again with Dirichlet data.

Shaw's proof of the above results is based on applying Schauder's fixed point theorem to the auxiliary equation, and on a remark, or topological lemma, which can be traced in Leray–Schauder: *If $Z$ is a compact convex subset of a Banach space, if $F$ is a continuous map from $[a, b] \times Z$ into $Z$, where $[a, b]$ is a closed interval, then there is a connected set of points $[(t, z) | t \in [a, b], z \in Z, F(t, z) = Z]$ that meets both $\{a\} \times Z$ and $\{b\} \times Z$.* In other words, while in the analysis based on contraction maps, we know that the unique solution $\mathfrak{X}(t)$ of the auxiliary equation is a continuous function of $t$, here with this remark, a kind of "continuity" is established in the set-valued map $\mathfrak{X}(t)$ obtained by the use of Schauder's fixed point theorem on the auxiliary equation.

We come now to the case of multiple resonance, say the case where $E$ has Fredholm indices $(k, k)$, $k > 1$. Shaw has found ways to apply an algebraic topological argument to the map represented by the bifurcation equation. Here is the topological statement:

*Let $S$ be a compact topological $p$-manifold ($p \geqslant 1$, integer) with boundary $\partial S$, where $\partial S$ is homeomorphic to $\partial I^p =$ the $(p-1)$-sphere $= \{x \in R^p | \sup|x_i| = 1, i = 1, \ldots, p\}$. Assume that there is a continuous map $V : S \to I^p$ with $V : \partial S \to \partial I^p$, and that the topological degree of $V|_{\partial S}$ relative to the homeomorphic equivalence of $\partial S$ and $\partial I^p$ is $+1$ or $-1$. Then $V$ maps $S$ onto $I^p$.*

The proof is based on commutative diagrams and long exact sequences of cohomology groups. We refer to [65] for a precise statement of the ensuing existence theorem.

## REFERENCES

1. H. Brézis, *Opérateurs Maximaux Monotones*, American Elsevier, New York, 1973.

2. H. Brézis, M. Crandall, and A. Pazy, "Perturbation of nonlinear maximal monotone sets," *Comm. Pure Appl. Math.*, **23** (1970), 123–144.

3. L. Cesari, "Functional analysis and periodic solutions of nonlinear differential equations," *Contributions to Differential Equations 1*, Wiley, New York, 1963, 149–187.

4. ———, "Functional analysis and Galerkin's method," *Michigan Math. J.*, **11** (1964), 385–414.

5. ———, "Existence in the large of periodic solutions of hyperbolic partial differential equations," *Arch. Rational Mech. Anal.*, **20** (1965), 170–190.

6. ———, "A nonlinear problem in potential theory," *Michigan Math. J.*, **16** (1969), 3–20.

7. ———, "Alternative methods in nonlinear analysis," *International Conference on Differential Equations* (H. A. Antosiewicz, *ed.*), Academic Press, New York, 1975, 95–148.

8. ———, "Nonlinear oscillations in the frame of alternative methods," *International Conference on Dynamical Systems,* Providence, R.I., vol. 1, Academic Press, New York, 1976, 29–50.

9. ———, "Functional analysis, nonlinear differential equations, and the alternative method," *Nonlinear Functional Analysis and Differential Equations* (L. Cesari, R. Kannan, and J. Schuur, *eds.*), M. Dekker, New York, 1976, 1–197.

10. L. Cesari and T. T. Bowman, "Some error estimates by the alternative method," *Quart. Appl. Math.*, (to appear).

11. L. Cesari and R. Kannan, "Functional analysis and nonlinear differential equations," *Bull. Amer. Math. Soc.*, **79** (1973), 1216–1219.

12. ———, "Periodic solutions in the large of nonlinear ordinary differential equations," *Rend. Mat., Università di Roma*, (2) **8** (1975), 633–654.

13. ———, "Solutions in the large of Liénard systems with forcing terms," *Ann. Mat. Pura Appl.*, (to appear).

14. D. G. De Figueiredo, "Some remarks on the Dirichlet problem for semilinear elliptic equations," *Univ. de Brasilia, Trabalho de Mat.*, **57** (1974).

15. ———, "On the range of nonlinear operators with linear asymptotes which are not invertible," *Univ. de Brasilia, Trabalho de Mat.*, **59** (1974).

16. ———, "The Dirichlet problem for nonlinear elliptic equations: a Hilbert space approach," *Partial Differential Equations and Related Topics* (Dold and Eckman ed.), Springer Verlag Lecture Notes in Mathematics, **446** (1975), 144–165.

17. S. Fucik, "Further remarks on a theorem by E. Landesman and A. C. Lazer," *Comment. Math. Univ. Carolin.*, **15** (1974), 259–271.

18. S. Fucik, M. Kucera, and J. Necas, "Ranges of nonlinear asymptotically linear operators," *J. Differential Equations*, **17** (1975), 375–394.

19. D. Graffi, "Sulle oscillazioni forzate nella meccanica nonlineare," *Riv. Mat. Univ. Parma*, **3** (1952), 317–326.

20. ———, "Forced oscillations for several nonlinear circuits," *Ann. of Math.*, **54** (1951), 262–271.

21. K. Gustafson and D. Sather, "Large nonlinearities and monotonicity," *Arch. Rational Mech. Anal.*, **48** (1972), 109–122.

22. ———, "Large nonlinearities and closed linear operators," *Arch. Rational Mech. Anal.*, **52** (1973), 10–19.

23. J. K. Hale, "Applications of alternative problems," *Lecture Notes*, Brown University, 1971.

24. ———, "Periodic solutions of a class of hyperbolic equations containing a small parameter," *Arch. Rational Mech. Anal.*, **23** (1967), 380–398.

25. ———, *Ordinary Differential Equations*, Interscience, 1969.

26. J. K. Hale, S. Bancroft, and D. Sweet, "Alternative problems for nonlinear equations," *J. Differential Equations*, **4** (1968), 40–56.

27. P. Hess, "On a theorem by Landesman and Lazer," *Indiana Univ. Math. J.*, **23** (1974), 827–829.

28. R. Kannan, "Periodically disturbed conservative systems," *J. Differential Equations*, **16** (1974), 506–514.

29. ———, "Existence of periodic solutions of differential equations," *Trans. Amer. Math. Soc.*, **217** (1976), 225–236.

30. ———, "Existence of solutions of a nonlinear problem in potential theory," *Michigan Math. J.*, **21** (1974), 257–264.

31. R. Kannan and J. Locker, "Operators $J^*J$ and nonlinear Hammerstein equations," *J. Math. Anal. Appl.*, **53** (1976), 1–7.

32. R. Kannan and H. Salehi, "Random solutions of nonlinear differential equations," *Boll. Un. Mat. Ital.*, (4) **12** (1975), 209–213.

33. T. Kato, "Demicontinuity, hemicontinuity, and monotonicity, I and II," *Bull. Amer. Math. Soc.*, **73** (1967), 470–476, 886–889.

34. H. W. Knobloch, "Remarks on a paper of Cesari on functional analysis and nonlinear differential equations," *Michigan Math. J.*, **10** (1963), 417–430.

35. ———, "Eine neue Methode zur Approximation von periodischen Lösungen nicht linear Differentialgleichungen zweiter Ordnung," *Math. Z.*, **82** (1963), 177–197.

36. D. Ku, *Boundary Value Problems and Numerical Estimates*, University of Michigan, thesis, 1976.

37. A. C. Lazer, "On Schauder's fixed point theorem and forced second order nonlinear oscillations," *J. Math. Anal. Appl.*, **21** (1968), 421–425.

38. E. M. Landesman and A. Lazer, "Nonlinear perturbations of linear elliptic boundary value problems at resonance," *J. Math. Mech.*, **19** (1970), 609–623.

39. S. Lefschetz, "Existence of periodic solutions of certain differential equations," *Proc. Nat. Acad. Sci.*, **29** (1943), 29–32.

40. J. Leray and J. Schauder, "Topologie et équations fonctionnelles," *Ann. Sci. École Norm. Sup.*, **51** (1934), 45–78.

41. N. Levinson, "On the existence of periodic solutions for second order differential equations with a forcing term," *J. Mathematical Phys.*, **22** (1943), 41–48.

42. J. Locker, "An existence analysis for nonlinear equations in Hilbert spaces," *Trans. Amer. Math. Soc.*, **128** (1967), 403–413.

43. A. M. Lyapunov, "Sur les figures d'équilibre peu différentes des ellipsoïdes d'une masse liquide homogène douée d'un mouvement de rotation," *Zap. Akad. Nauk, St. Petersburg*, **1** (1906), 1–225.

44. J. Mawhin, "Le problème des solutions périodiques en mécanique non linéaire," *Thèse, Université de Liège*, 1969.

45. ———, "Periodic solutions of nonlinear functional differential equations," *J. Differential Equations*, **10** (1971), 240–261.

46. ———, "Degré topologique et solutions périodiques des systèmes différentiels nonlinéaires," *Bull. Soc. Roy. Sci. Liège*, **38** (1969), 308–398.

47. ———, "Periodic solutions of strongly nonlinear differential systems," *Proc. Fifth International Congress on Nonlinear Oscillations, Kiev*, **1** (1969), 380–399.

48. ———, "An extension of a theorem of A. C. Lazer on forced nonlinear oscillations," *J. Math. Anal. Appl.*, **40** (1972), 20–29.

49. ———, "Equivalence theorems for nonlinear operator equations and coincidence degree theory for some mappings in locally convex topological vector spaces," *J. Differential Equations*, **12** (1972), 610–636.

50. ———, "Topology and nonlinear boundary value problems," *International Conference on Dynamical Systems*, Providence, R.I., vol. 1, Academic Press, New York, 1976, 51–82.

51. ———, "$L_2$-estimates and periodic solutions of some nonlinear differential equations," *Boll. Un. Mat. Ital.*, (4) **10** (1974), 341–352.

52. ———, "Nonlinear perturbations of Fredholm mappings in normed spaces and applications to differential equations," *Univ. de Brasilia, Trabalho de Mat.*, **58** (1974).

53. G. Minty, "On a monotonicity method for the solution of nonlinear equations in Banach spaces," *Proc. Nat. Acad. Sci. U.S.A.*, **50** (1963), 1038–1041.

54. S. Mizohata, "On the existence of periodic solutions for several nonlinear circuits," *Mem. Coll. Sci., University of Kyoto, Ser. A Math.*, **27** (1952), 115–121.

55. S. Mizohata and M. Yamaguti, "On the existence of periodic solutions of a second order nonlinear differential equation," *Mem. Coll. Sci. University of Kyoto, Ser. A Math.*, **27** (1951), 109–113.

56. K. Nagle, *Boundary Value Problems for Nonlinear Ordinary Differential Equations*, University of Michigan, thesis, 1975.

57. M. Nagumo, "A theory of degree of mappings based on infinitesimal analysis," *Amer. J. Math.*, **73** (1951), 485–496.

58. ———, "Degree of mappings in convex linear topological spaces," *Amer. J. Math.*, **73** (1951), 497–511.

59. J. Necas, "On the range of nonlinear operators with linear asymptotes which are not invertible," *Comm. Math. Univ. Carolin.*, **14** (1973), 63–72.

60. J. E. Osborn and D. Sather, "Alternative problems for nonlinear equations," *J. Differential Equations*, **17** (1975), 12–31.

61. ———, "Alternative problems and monotonicity," *J. Differential Equations*, **18** (1975), 393–410.

62. H. Poincaré, *Les méthodes nouvelles de la mécanique céleste*, Gauthier-Villars, Paris, 1892–1899. Dover, New York, 1957.

63. D. A. Sanchez, "An iteration scheme for Hilbert space boundary value problems," *Boll. Un. Mat. Ital.*, (4) **11** (1975), 1–9.

64. E. Schmidt, "Zur Theorie der linearen und nichtlinearen Integralgleichungen und der Verzweigung ihrer Lösungen," *Math. Ann.*, **65** (1908), 370–399.

65. H. C. Shaw, *Nonlinear Elliptic Boundary Value Problems at Resonance*, University of Michigan, thesis, 1975.

66. S. A. Williams, "A connection between the Cesari and Leray–Schauder methods," *Michigan Math. J.*, **15** (1968), 441–448.

67. ———, "A sharp sufficient condition for solution of a nonlinear elliptic boundary value problem," *J. Differential Equations*, **8** (1970), 580–586.

# ASYMPTOTIC METHODS

*Yasutaka Sibuya*[†]

## I. SUCCESSIVE PARTIAL SUMS OF A SERIES AS SUCCESSIVE APPROXIMATIONS*

**1. Examples.** Let us consider two series

$$\sum_{n=0}^{\infty} \frac{(1000)^n}{n!} \tag{1.1}$$

and

$$\sum_{n=0}^{\infty} \frac{n!}{(1000)^n}\,. \tag{1.2}$$

†Partially supported by the National Science Foundation under Grant No. GP-38955.

The author would like to thank Dr. Ivar Bakken who read the manuscript and gave many valuable comments.

*Cf. H. Jeffreys [12; Chapter 1].

Apparently, series (1.1) is convergent, but series (1.2) is divergent. H. Poincaré, however, pointed out that the terms in series (1.1) increase up to the 1000th, while the terms in series (1.2) decrease up to the 1000th.* Suppose that we want to get an approximation of a quantity within a reasonable error-bound. A reasonable error means an error that is sufficiently small. It is not necessary to obtain an arbitrarily small error. Suppose that successive partial sums of a series are utilized as successive approximations. In such a case, the rapid decrease of the early terms is frequently found desirable more than the convergence of the series itself. When this point of view is adopted, the convergence and the divergence of series become relatively insignificant. The center of interest is the behavior of successive sums. Thus begins a story of "Asymptotic expansions".

For a further illustration, let us consider an alternating series

$$\sum_{n=0}^{\infty} (-1)^n a_n, \tag{1.3}$$

where the quantities $a_n$ are all positive. Suppose that partial sums

$$S_N = \sum_{n=0}^{N} (-1)^n a_n \qquad (N = 0, 1, \ldots)$$

of series (1.3) are utilized as successive approximations of a quantity $Q$. The error of the $N$th approximation is given by

$$E_N = Q - S_N.$$

Set

$$R_N = |E_N| \qquad (N = 0, 1, \ldots).$$

Suppose that the errors $E_N$ also alternate in sign, i.e.,

$$E_N = (-1)^{N+1} R_N \qquad (N = 0, 1, \ldots). \tag{1.4}$$

*H. Poincaré [22; vol. II, Chapter VIII].

Observe that

$$Q = S_N + E_N = S_{N+1} + E_{N+1}.$$

Then, it follows that

$$E_N = S_{N+1} - S_N + E_{N+1},$$

or

$$(-1)^{N+1} R_N = (-1)^{N+1} a_{N+1} + (-1)^{N+2} R_{N+1}.$$

Hence

$$R_N = a_{N+1} - R_{N+1} \leqslant a_{N+1},$$

or

$$|E_N| \leqslant a_{N+1}.$$

This means that the absolute value of the error of the $N$th approximation (i.e., $R_N$) is not larger than the absolute value of the $(N + 1)$st term of series (1.3) (i.e., $a_{N+1}$). Therefore, if at least one term of series (1.3) becomes reasonably small in the absolute value, we will get a good approximation of the quantity $Q$. The convergence or the divergence of series (1.3) is not important. The most important thing to do is to stop calculation at the right place. It is not only wasteful, but also harmful, to calculate too many terms.*

**2. An example of the example.** This entire story might sound like a dubious myth. But, here is a more legitimate case. It is not difficult to verify that the function

$$f(t) = -e^t \int_t^{+\infty} s^{-1} e^{-s} \, ds \tag{2.1}$$

*Cf. G. G. Stokes [49; §7, p. 171].

is a solution of the differential equation

$$y' - y = t^{-1} \qquad (y' = dy/dt). \tag{2.2}$$

A repeated integration by parts yields

$$\begin{aligned} f(t) &= -t^{-1} + e^t \int_t^{+\infty} s^{-2} e^{-s}\, ds \\ &= -t^{-1} + t^{-2} - 2e^t \int_t^{+\infty} s^{-3} e^{-s}\, ds \\ &= -t^{-1} + t^{-2} - 2t^{-3} + 6e^t \int_t^{+\infty} s^{-4} e^{-s}\, ds \\ &= -\left\{ \sum_{n=0}^{N} (-1)^n \frac{n!}{t^{n+1}} \right. \\ &\qquad \left. + (-1)^{N+1} (N+1)!\, e^t \int_t^{+\infty} s^{-N-2} e^{-s}\, ds \right\}, \end{aligned}$$

or

$$Q = \sum_{n=0}^{N} (-1)^n a_n + (-1)^{N+1} R_N,$$

where

$$Q = -f(t), \quad a_n = \frac{n!}{t^{n+1}}, \quad R_N = (N+1)!\, e^t \int_t^{+\infty} s^{-N-2} e^{-s}\, ds.$$

Observe that, for $t > 0$, we have

$$a_n > 0 \quad \text{and} \quad R_N > 0.$$

This means that all of the assumptions concerning series (1.3) are satisfied. In particular, assumption (1.4) is satisfied. Hence

$$R_N < a_{N+1} \quad \text{for} \quad t > 0,$$

or

$$\left| f(t) - \sum_{n=0}^{N} (-1)^{n+1} \frac{n!}{t^{n+1}} \right| < \frac{(N+1)!}{t^{N+2}} \quad \text{for} \quad t > 0. \tag{2.3}$$

The power series

$$\sum_{n=0}^{\infty} (-1)^{n+1} \frac{n!}{t^{n+1}} \tag{2.4}$$

is divergent. Nevertheless, its partial sums may be used as approximations of $f(t)$. Inequality (2.3) provides us with an estimate of the error of such an approximation. Power series (2.4) is called the asymptotic expansion of $f(t)$ in the sense of Poincaré as $t$ tends to $+\infty$. The concept of asymptotic expansions will be explained more precisely in Section 4.

Power series (2.4) was derived from particular solution (2.1) of differential equation (2.2). However, it is possible to derive power series (2.4) directly from differential equation (2.2). The idea is similar to the method of undetermined coefficients. To do this, set

$$y = \sum_{n=0}^{\infty} A_n t^{-n} \quad \text{and} \quad y' = \sum_{n=0}^{\infty} (-n) A_n t^{-n-1}.$$

Then,

$$y' - y = A_0 + \sum_{n=0}^{\infty} \left[ -(n-1)A_{n-1} - A_n \right] t^{-n}.$$

Since $y' - y$ must be equal to $t^{-1}$, let us set

$$\begin{cases} A_0 = 0 \quad , \\ -A_1 = 1 \quad , \\ -(n-1)A_{n-1} - A_n = 0 \qquad (n \geqslant 2). \end{cases}$$

Then, it follows that

$$A_0 = 0, \quad A_n = (-1)^n (n-1)! \qquad (n \geqslant 1),$$

or

$$\sum_{n=0}^{\infty} A_n t^{-n} = \sum_{n=0}^{\infty} (-1)^{n+1} \frac{n!}{t^{n+1}}.$$

Power series (2.4) is called a formal solution of differential equation (2.2). The concept of formal solutions will be explained more precisely in Section 3.

In general, there are two problems concerning differential equations.

PROBLEM I: *For a given solution, find its asymptotic expansions.*

PROBLEM II: *For a given formal solution, find a (genuine) solution so that the given formal solution is an asymptotic expansion of this solution.*

In this section, we solved Problem I for particular solution (2.1) of differential equation (2.2) as $t$ tends to $+\infty$.

The concept of asymptotic expansions probably had sprouted and been nursed in the practical world. Then, eventually, this concept was found legitimate even in the theoretical world. In history, however, there are some evidences of protest against it. Poincaré utilized his asymptotic method in his study of linear differential equations at irregular singular points.* Apparently, "M. Thomé" complained, and Poincaré found it necessary to defend his method.**

## II. ALGEBRAIC DIFFERENTIAL EQUATIONS

**3. Formal solutions.** We called power series (2.4) a formal solution of differential equation (2.2). In this section, the concept of formal solutions will be explained more precisely.

---

*H. Poincare [44 and 45].

**H. Poincaré [46].

Corresponding to any ordered denumerable set $\{a_0, a_1, \ldots, a_m, \ldots\}$ of complex numbers, we form the expression $\Sigma_{m=0}^{\infty} a_m t^{-m}$, and we call it a formal power series in $t^{-1}$ over the complex field **C**. Let us denote by $\mathbf{C}[[t^{-1}]]$ the set of all formal power series in $t^{-1}$ over **C**.

For two elements $p = \Sigma_{m=0}^{\infty} a_m t^{-m}$ and $q = \Sigma_{m=0}^{\infty} b_m t^{-m}$ of $\mathbf{C}[[t^{-1}]]$,

$$p = q \quad \text{means} \quad a_m = b_m \quad \text{for all} \quad m. \tag{3.1}$$

Let $\lambda$ be a complex number. Set

$$p + q = \sum_{m=0}^{\infty} (a_m + b_m) t^{-m}, \quad \lambda p = \sum_{m=0}^{\infty} \lambda a_m t^{-m}. \tag{3.2}$$

Then, $\mathbf{C}[[t^{-1}]]$ is a vector space over **C** with operations (3.2). In particular, $p = \Sigma_{m=0}^{\infty} a_m t^{-m}$ is the zero-element $\mathbf{0}$ when all $a_m$ are zero.

Set also

$$pq = \sum_{m=0}^{\infty} \Big( \sum_{h+k=m} a_h b_k \Big) t^{-m}. \tag{3.3}$$

Then, $\mathbf{C}[[t^{-1}]]$ is a commutative ring with the operations $p + q$ and $pq$. This ring has the unit-element $\mathbf{1}$, and does not have zero-divisors. The unit-element $\mathbf{1}$ is given by $\mathbf{1} = \Sigma_{m=0}^{\infty} \delta_{0,m} t^{-m}$, where $\delta_{0,0} = 1$, and $\delta_{0,m} = 0$ $(m \geqslant 1)$. With the operations $p + q$, $pq$, and $\lambda p$, the set $\mathbf{C}[[t^{-1}]]$ is a commutative algebra with the unit-element, over **C**.*

Set

$$p' = \sum_{m=0}^{\infty} (-m) a_m t^{-m-1}. \tag{3.4}$$

The element $p'$ of $\mathbf{C}[[t^{-1}]]$ is called the first derivative of the element $p = \Sigma_{m=0}^{\infty} a_m t^{-m}$. Successively, we can define the $k$th

*Cf. R. J. Walker [25; Chapter IV], and H. Cartan [6; Chapter I].

derivative $p^{(k)}$ of $p$. It is evident that $(p+q)' = p' + q'$ and $(pq)' = p'q + pq'$. Note that $p^{(k)} = \mathbf{0}$ for some $k$ if and only if $p = \lambda\mathbf{1}$ for some complex number $\lambda$.

An element $p = \sum_{m=0}^{\infty} a_m t^{-m}$ of $\mathbf{C}[[t^{-1}]]$ is called a polynomial in $t^{-1}$ over $\mathbf{C}$, when $a_m = 0$ except for a finite number of them. We denote by $\mathbf{C}[t^{-1}]$ the set of all polynomials in $t^{-1}$ over $\mathbf{C}$. The set $\mathbf{C}[t^{-1}]$ is a subalgebra of $\mathbf{C}[[t^{-1}]]$. The algebra $\mathbf{C}[[t^{-1}]]$ may be regarded as a differential algebra over $\mathbf{C}$, or $\mathbf{C}[t^{-1}]$, or $\mathbf{C}[[t^{-1}]]$ itself. The algebra $\mathbf{C}[t^{-1}]$ is a subalgebra of $\mathbf{C}[[t^{-1}]]$ as a differential algebra over $\mathbf{C}$, or $\mathbf{C}[t^{-1}]$.*

Let $F(\xi, x_0, x_1, \ldots, x_n)$ be a polynomial in $\xi, x_0, \ldots, x_n$ over $\mathbf{C}$, i.e.,

$$F(\xi, x) = \sum_{l+l_0+\cdots+l_n \leq N} c_{ll_0l_1\cdots l_n} \xi^l x_0^{l_0} x_1^{l_1} \cdots x_n^{l_n},$$

where $c_{ll_0l_1\cdots l_n} \in \mathbf{C}$, and $l, l_0, l_1, \ldots, l_n$ are non-negative integers. Then, a differential equation

$$F(t^{-1}, y, y', \ldots, y^{(n)}) = 0 \tag{3.5}$$

is called an algebraic differential equation, where $y$ is the unknown quantity, and $t$ is the independent variable. The polynomial $F(t^{-1}, x)$ may be written also in the form

$$F(t^{-1}, x) = \sum_{l_0+l_1+\cdots+l_n \leq N} p_{l_0l_1\cdots l_n} x_0^{l_0} x_1^{l_1} \cdots x_n^{l_n},$$

where $p_{l_0l_1\cdots l_n} \in \mathbf{C}[t^{-1}]$. Therefore, if $p \in \mathbf{C}[[t^{-1}]]$, we have $F(t^{-1}, p, p', \ldots, p^{(n)}) \in \mathbf{C}[[t^{-1}]]$. Furthermore,

$$F(t^{-1}, p, p', \ldots, p^{(n)}) = \mathbf{0} \tag{3.6}$$

is a legitimate equation in the algebra $\mathbf{C}[[t^{-1}]]$. Solutions of equation (3.6) are called formal solutions of differential equation (3.5).

*For differential algebras, see E. R. Kolchin [14].

Let us consider a particular case when $F$ is linear and homogeneous in $y, \ldots, y^{(n)}$, i.e., $F(t^{-1}, y, \ldots, y^{(n)}) = \Sigma_{k=0}^{n} p_k y^{(k)}$, where $p_k \in \mathbf{C}[t^{-1}]$. Set $L(p) = F(t^{-1}, p, p', \ldots, p^{(n)})$ for $p \in \mathbf{C}[[t^{-1}]]$. Then, $L(p)$ is a linear operator defined on $\mathbf{C}[[t^{-1}]]$, where $\mathbf{C}[[t^{-1}]]$ is regarded as a vector space over $\mathbf{C}$. Denote by $\mathbf{C}\langle\langle t^{-1}\rangle\rangle$ the set of all convergent power series in $t^{-1}$ over $\mathbf{C}$. The set $\mathbf{C}\langle\langle t^{-1}\rangle\rangle$ is a subspace of $\mathbf{C}[[t^{-1}]]$ as a vector space over $\mathbf{C}$. Let $K(L)$ be the set of all elements $p$ of $\mathbf{C}[[t^{-1}]]$ such that $L(p)$ is a convergent power series in $t^{-1}$ (i.e., $L(p) \in \mathbf{C}\langle\langle t^{-1}\rangle\rangle$). It is evident that $K(L)$ is a subspace of $\mathbf{C}[[t^{-1}]]$ and that $\mathbf{C}\langle\langle t^{-1}\rangle\rangle$ is a subspace of $K(L)$. Denote by $i(L)$ the codimension of $\mathbf{C}\langle\langle t^{-1}\rangle\rangle$ as a subspace of $K(L)$. The number $i(L)$ is called the irregularity of $L$ at $t = \infty$. Set $p_k = t^{-m_k}(c_{k,0} + c_{k,1}t^{-1} + \cdots)$, where $c_{k,0} \neq 0$. Then,

$$i(L) = (n + m_n) - \min_{k=0}^{n} (k + m_k).^* \tag{3.7}$$

When $i(L) = 0$, the point $t = \infty$ is called (at most) a regular singular point of $L$. Otherwise, $t = \infty$ is called an irregular singular point of $L$.** For example, if we put $L(p) = p' - p$, and if we set $p_0 = \Sigma_{n=0}^{\infty}(-1)^{n+1}(n!)/(t^{n+1})$, then, $L(p_0) = t^{-1} \in \mathbf{C}[t^{-1}] \subset \mathbf{C}\langle\langle t^{-1}\rangle\rangle$. Since $p_0$ is not convergent, but $p_0 \in K(L)$, the point $t = \infty$ is an irregular singular point of $L$. As a matter of fact, $i(L) = 1$ in this case. The condition "$i(L) = 0$" means that $L(p) \in \mathbf{C}\langle\langle t^{-1}\rangle\rangle$ implies $p \in \mathbf{C}\langle\langle t^{-1}\rangle\rangle$. Therefore, if $t = \infty$ is a regular singular point of $L$, all formal solutions are convergent.***

Generally speaking, it is extremely difficult to find formal solutions.**** Furthermore, many algebraic differential equations

---

*Cf. B. Malgrange [41; pp. 149–150].

**For the classical definitions of regular and irregular singularities, see E. A. Coddington and N. Levinson [8; Chapters 4 and 5].

***Cf. W. Wasow [26; Chapter II, Theorem 5.3, p. 22].

****A book of A. H. Nayfeh provides us with a collection of fine arts of formal calculation in the practical world. (Cf. A. H. Nayfeh [16].) See also H. Poincaré [22; vol. II].

do not admit formal solutions of this kind. For such differential equations, we must consider much larger algebras containing $\mathbf{C}[[t^{-1}]]$. For example, in case when $F$ is linear and homogeneous in $y, y', \ldots, y^{(n)}$, and when $t = \infty$ is a regular singular point, differential equation (3.5) admits formal solutions of the form $t^{\alpha}S(t^{-1}, \log t)$, where $\alpha$ is a complex number, and $S$ is a polynomial in $\log t$ with coefficients in $\mathbf{C}[[t^{-1}]]$, i.e., $S(t^{-1}, \log t) = \sum_{h=0}^{N}(\log t)^h p_h$, and $p_h \in \mathbf{C}[[t^{-1}]]$.* If $t = \infty$ is an irregular singular point, differential equation (3.5) admits formal solutions of the form $e^{P(t)}t^{\alpha}S(t^{-1/k}, \log t)$, where $k$ is a positive integer, $\alpha$ is a complex number, $P(t)$ is a polynomial in $t^{1/k}$, and $S$ is a polynomial in $\log t$ with coefficients in $\mathbf{C}[[t^{-1/k}]]$.**

**4. Asymptotic expansions.** We called power series (2.4) the asymptotic expansion of function (2.1) in the sense of Poincaré as $t$ tends to $+\infty$. In this section, the concept of asymptotic expansions will be explained more precisely.

Let $\mathfrak{S}$ be a connected unbounded set in the complex $t$-plane, and let $f(t)$ be a function defined in $\mathfrak{S}$. An element $p = \sum_{m=0}^{\infty} a_m t^{-m}$ of $\mathbf{C}[[t^{-1}]]$ is said to be an asymptotic expansion of $f$ (in the sense of Poincaré) as $t$ tends to infinity in $\mathfrak{S}$, when

$$\left| f(t) - \sum_{m=0}^{N} a_m t^{-m} \right| \leqslant K_N |t|^{-N-1} \quad \text{for} \quad t \in \mathfrak{S},$$

$$(N = 0, 1, \ldots), \tag{4.1}$$

where the quantities $K_N$ are positive constants independent of $t$. The asymptotic expansion $p$ of $f$ in $\mathfrak{S}$ is uniquely determined by $f$ if it exists. In fact,

$$\begin{cases} a_0 = \lim f(t), \\ a_N = \lim t^N \left\{ f(t) - \sum_{m=0}^{N-1} a_m t^{-m} \right\} \quad (N \geqslant 1), \end{cases} \tag{4.2}$$

*E. A. Coddington and N. Levinson [8; Chapter 4, §§3, 7, and 8].

**E. A. Coddington and N. Levinson [8; Chapter 5, §6, Theorem 6.1].

as $t$ tends to infinity in $\mathbb{S}$. In the case of power series (2.4) and function (2.1), the set $\mathbb{S}$ is the half-line $0 < t < +\infty$.

Let us denote by $A(\mathbb{S})$ the set of all functions $f$ such that

(1) $f$ is defined in $\mathbb{S}$, and

(2) there exists an element $p$ in $\mathbf{C}[[t^{-1}]]$ which is an asymptotic expansion of $f$ as $t$ tends to infinity in $\mathbb{S}$.

For $f \in A(\mathbb{S})$, its unique asymptotic expansion is denoted by $A_{\text{sp}}(f)$. Then, $A_{\text{sp}}(f)$ is a mapping of $A(\mathbb{S})$ into $\mathbf{C}[[t^{-1}]]$.

For two elements $f$ and $g$ of $A(\mathbb{S})$, and for a complex number $\lambda$, we set

$$(f+g)(t) = f(t) + g(t), \quad (fg)(t) = f(t)g(t), \quad (\lambda f)(t) = \lambda f(t).$$

Then, $f + g$, $fg$, and $\lambda f$ are also elements of $A(\mathbb{S})$, and

$$\begin{cases} A_{\text{sp}}(f+g) = A_{\text{sp}}(f) + A_{\text{sp}}(g), \quad A_{\text{sp}}(fg) = A_{\text{sp}}(f)A_{\text{sp}}(g), \\ A_{\text{sp}}(\lambda f) = \lambda\, A_{\text{sp}}(f). \end{cases}$$

In other words, $A(\mathbb{S})$ is a commutative algebra with the unit-element, over $\mathbf{C}$, and $A_{\text{sp}}(f)$ is a homomorphism of $A(\mathbb{S})$ into $\mathbf{C}[[t^{-1}]]$.*

The unit-element $\mathbf{1}(t)$ and the zero-element $\mathbf{0}(t)$ of $A(\mathbb{S})$ are given by

$$\begin{cases} \mathbf{1}(t) = 1 \quad \text{for all} \quad t \text{ in } \mathbb{S}, \\ \mathbf{0}(t) = 0 \quad \text{for all} \quad t \text{ in } \mathbb{S}, \end{cases}$$

respectively. In general, for a polynomial $p = \sum_{m=0}^{\infty} a_m t^{-m}$ over $\mathbf{C}$, we set

$$p(t) = a_0 + a_1 t^{-1} + a_2 t^{-2} + \cdots \quad \text{for all values of} \quad t.$$

Note that the right-hand member is a sum of a finite number of

*Cf. E. A. Coddington and N. Levinson [8; Chapter 5, §3]. See also W. Wasow [26; Chapter III, §§7 and 8], and L. Sirovich [23; Chapter 1, §§1.1 and 1.2].

terms. In this way, we can regard the algebra $\mathbf{C}[t^{-1}]$ as a subalgebra of $A(\mathfrak{S})$, and $A_{\text{sp}}(p) = p$ for all $p \in \mathbf{C}[t^{-1}]$.

Generally speaking, $A_{\text{sp}}(f)$ *does not have* the following three properties:

(I) $A_{\text{sp}}(f)$ is injective (i.e., one-to-one);

(II) $A_{\text{sp}}(f)$ is surjective (i.e., onto);

(III) $A_{\text{sp}}(f') = (A_{\text{sp}}(f))'$.

In property (III), we must assume that an element $f$ of $A(\mathfrak{S})$ admits the first derivative $f'$ with respect to $t$, and $f' \in A(\mathfrak{S})$.

Let $\mathfrak{B}$ be a subalgebra of $A(\mathfrak{S})$ such that

(i) for every $f \in \mathfrak{B}$, the first derivative $f'$ exists, and $f' \in \mathfrak{B}$,

(ii) $\mathbf{C}[t^{-1}]$ is a subalgebra of $\mathfrak{B}$, and

(iii) the mapping $A_{\text{sp}}(f)$ which is restricted on $\mathfrak{B}$ satisfies conditions (I) and (III).

Condition (I) in this case means that, if $f \in \mathfrak{B}$ and $A_{\text{sp}}(f) = \mathbf{0}$, then $f = \mathbf{0}$. Every function $f$ in $\mathfrak{B}$ is completely determined by quantities (4.2). Denote by $A_{\text{sp}}(\mathfrak{B})$ the image of $\mathfrak{B}$ by the mapping $A_{\text{sp}}(f)$. It is evident that $A_{\text{sp}}(\mathfrak{B})$ is a subalgebra of $\mathbf{C}[[t^{-1}]]$. Furthermore, the mapping $A_{\text{sp}}(f)$ is an isomorphism of $\mathfrak{B}$ onto $A_{\text{sp}}(\mathfrak{B})$. Assume that an element $p$ of $A_{\text{sp}}(\mathfrak{B})$ is a formal solution of differential equation (3.5). Let $f$ be the unique element of $\mathfrak{B}$ such that $A_{\text{sp}}(f) = p$. Then,

$$A_{\text{sp}}\big(F(t^{-1}, f, f', \ldots, f^{(n)})\big) = F(t^{-1}, p, p', \ldots, p^{(n)}) = \mathbf{0}. \quad (4.3)$$

Since $F(t^{-1}, f, f', \ldots, f^{(n)}) \in \mathfrak{B}$, we have $F(t^{-1}, f, f', \ldots, f^{(n)}) = \mathbf{0}$. In other words, $y = f(t)$ is a solution of differential equation (3.5). Therefore, it is important to find a subalgebra $\mathfrak{B}$ of $A(\mathfrak{S})$ which satisfies conditions (i), (ii), and (iii). By using this guide-line, we can find some methods of summing formal power series. The requirement that a formal solution $p$ of differential equation (3.5) belongs to $A_{\text{sp}}(\mathfrak{B})$ is usually expressed as a summability condition on $p$ in a certain sense.* For example, if a formal solution $p$ of (3.5) is convergent in the usual sense, its sum satisfies differential equation (3.5).

---

*Cf. E. Borel [3; Chapters II and III; in particular, pp. 84 and 148].

As to condition (I) on the mapping $A_{sp}(f)$, the following criterion* is very important. Assume that $f(t)$ is holomorphic in the half-plane $\mathrm{Re}(t) \geqslant r_0 > 0$. Then, the condition that

$$|f(t)| \leqslant \left| \frac{C_N}{t} \right|^N \qquad (N = 0, 1, \ldots)$$

in this half-plane (the quantities $C_N$ being positive numbers) implies that $f(t)$ is identically equal to zero, if and only if the quantity

$$\int_1^{+\infty} \log\left[ \sum_{N=0}^{\infty} \left( \frac{r}{C_N} \right)^N \right] \frac{dr}{r^2}$$

diverges. In particular, if the series

$$\sum_{N=0}^{\infty} \frac{1}{C_N}$$

diverges, $f(t)$ is identically zero. Through an application of Laplace transform, this criterion can be carried over to the family of functions which admit all derivatives on the half-line $0 < t < +\infty$.**

In general, a formal solution $p$ *does not determine* a solution $y = f(t)$ uniquely by the condition $A_{sp}(f) = p$. Even when such a solution $f$ is uniquely determined by an additional condition $f \in \mathfrak{B}$, there might be more solutions satisfying the condition $A_{sp}(f) = p$.

As we mentioned in Section 3, there are formal solutions of more general forms. Corresponding to those general formal solutions, the concept of asymptotic expansions also must be generalized.***

---

*Cf. T. Carleman [5; Chapter V, p. 54], and F. Pittnauer [21; Chapter III].

**Cf. T. Carleman [5; Chapter VI, p. 61].

***Cf. L. Sirovich [23; Chapter 1].

**5. Asymptotic expansions in a sector.*** Let us consider a case when $\mathfrak{S}$ is a sector, i.e.,

$$\mathfrak{S} = \{t;\ a \leqslant \arg t \leqslant b,\ |t| \geqslant R\}, \tag{5.1}$$

where $a$, $b$, and $R$ are real numbers, $a < b$, and $R > 0$. In this case, the mapping $A_{\mathrm{sp}}(f)$ is surjective. In fact, for a given $p$ in $\mathbf{C}[[t^{-1}]]$, there exists a function $f(t)$ such that

($\alpha$) $f$ is holomorphic in a neighborhood of $\mathfrak{S}$, and

($\beta$) $A_{\mathrm{sp}}(f) = p$ as $t$ tends to infinity in $\mathfrak{S}$.

Furthermore, if $A_{\mathrm{sp}}(f) = \sum_{m=2}^{\infty} a_m t^{-m}$, we have

$$A_{\mathrm{sp}}\left(\int_t^{\infty} f(s)\,ds\right) = \sum_{m=1}^{\infty} \frac{a_{m+1}}{m}\, t^{-m}.$$

On the other hand, if $a < \hat{a} < \hat{b} < b$, conditions ($\alpha$) and ($\beta$) imply

($\gamma$) $A_{\mathrm{sp}}(f^{(k)}) = p^{(k)} \qquad (k = 1, 2, \ldots)$

as $t$ tends to infinity in the sector

$$\hat{\mathfrak{S}} = \{t;\ \hat{a} \leqslant \arg t \leqslant \hat{b},\ |t| \geqslant R\}. \tag{5.2}$$

Therefore, if $f^{(k)} \in A(\mathfrak{S})$, then ($\gamma$) holds as $t$ tends to infinity in $\mathfrak{S}$.

The mapping $A_{\mathrm{sp}}(f)$ *is not injective*. For example, if we set

$$f(t) = \exp\left[-\left(te^{-i\frac{1}{2}(a+b)}\right)^{\frac{\pi}{2(b-a)}}\right],$$

then $A_{\mathrm{sp}}(f) = \mathbf{0}$ in sector (5.1).

---

*Cf. W. Wasow [26; Chapter III].

## III. CONSTRUCTION OF A SOLUTION FOR A GIVEN FORMAL SOLUTION

**6. General.** Let us consider Problem II (cf. Section 2). Assume that an element $p$ of $\mathbf{C}[[t^{-1}]]$ is a formal solution of differential equation (3.5). Let $\mathbb{S}$ be a sector given by (5.1). Then, there exists a function $f(t)$ satisfying conditions ($\alpha$), ($\beta$), and ($\gamma$) of Section 5. In particular, $f$ satisfies condition ($\gamma$) as $t$ tends to infinity in $\mathbb{S}$. Therefore, (4.3) holds, i.e.,

$$A_{\mathrm{sp}}\big(F(t^{-1}, f, f', \dots, f^{(n)})\big) = \mathbf{0}. \tag{6.1}$$

Since the mapping $A_{\mathrm{sp}}(f)$ is not injective, the asymptotic result, such as (6.1), *does not imply* that $y = f(t)$ is a solution of (3.5).

Set $y = z + f(t)$. Then, differential equation (3.5) becomes

$$G(t, z, z', \dots, z^{(n)}) = h_0(t), \tag{6.2}$$

where

$$h_0(t) = -F\big(t^{-1}, f(t), f'(t), \dots, f^{(n)}(t)\big), \tag{6.3}$$

and

$$G(t, z, z', \dots, z^{(n)}) = F\big(t^{-1}, z + f(t), \dots, z^{(n)} + f^{(n)}(t)\big) + h_0(t).$$

The function $h_0(t)$ is holomorphic in $t$ in a neighborhood of $\mathbb{S}$, and $A_{\mathrm{sp}}(h_0) = \mathbf{0}$. (Cf. (6.3) and (6.1).) The function $G$ is a polynomial in $z, z', \dots, z^{(n)}$, such that $G(t, 0, 0, \dots, 0) = 0$. Furthermore, the coefficients of the polynomial $G$ are holomorphic in a neighborhood of $\mathbb{S}$, and they belong to $A(\mathbb{S})$.

Let us denote by $B(\mathbb{S})$ the set of all functions $g(t)$ satisfying the following conditions:

($\alpha'$) $g$ is continuous in $\mathbb{S}$, and holomorphic in the interior of $\mathbb{S}$;
($\beta'$) all of the derivatives $g^{(k)}$ are continuous in $\mathbb{S}$;
($\gamma'$) $g$ and $g^{(k)}$ ($k = 1, 2, \dots$) belong to $A(\mathbb{S})$; and
($\delta'$) $A_{\mathrm{sp}}(g) = \mathbf{0}$.

Conditions ($\gamma'$) and ($\delta'$) imply that $A_{sp}(g^{(k)}) = \mathbf{0}$. The set $B(\mathbb{S})$ is a subalgebra of $A(\mathbb{S})$ (without the unit-element). In particular, $B(\mathbb{S})$ is a subspace of $A(\mathbb{S})$ as a vector space over $\mathbf{C}$. Observe that $h_0 \in B(\mathbb{S})$. Furthermore, $G(t, g, g', \ldots, g^{(n)}) \in B(\mathbb{S})$ for all elements $g$ of $B(\mathbb{S})$. Therefore, differential equation (6.2) can be regarded as an equation in the space $B(\mathbb{S})$. If $z = g(t)$ is a solution of this equation in the space $B(\mathbb{S})$, then

$$y = g(t) + f(t) \tag{6.4}$$

is a solution of differential equation (3.5), and $A_{sp}(g + f) = A_{sp}(g) + A_{sp}(f) = \mathbf{0} + p = p$. Hence, (6.4) is a solution of Problem II concerning the formal solution $p$ of differential equation (3.5).

When we reduce differential equation (3.5) to (6.2), we can use a sector $\mathbb{S}$ of an arbitrary size. In other words, $\mathbb{S}$ may be as large as we want. However, in order to construct a solution of (6.2), we must replace $\mathbb{S}$ by a subsector $\hat{\mathbb{S}}$ of a reasonable size. Note that, if $\hat{\mathbb{S}}$ is a subsector of $\mathbb{S}$ which is given by (5.2), then $B(\mathbb{S})$ is a subspace of $B(\hat{\mathbb{S}})$. In general, an existence theorem for (6.2) is proved at most in a space $B(\hat{\mathbb{S}})$ for a suitable subsector $\hat{\mathbb{S}}$ of $\mathbb{S}$. The construction of such a subsector $\hat{\mathbb{S}}$ is frequently the center of the difficulty of Problem II.*

**7. A typical case.** It is difficult to solve equation (6.2) for general cases. In this section a typical result will be explained.

Let us consider the case when

$$G(t, z, z', \ldots, z^{(n)}) = q_n(t) z^{(n)} - H(t, z, z', \ldots, z^{(n-1)}),$$

where $q_n$ satisfies conditions ($\alpha'$), ($\beta'$), and ($\gamma'$) of Section 6. However, instead of condition ($\delta'$), we assume that

($\delta''$) $A_{sp}(q_n) \neq \mathbf{0}$.

The function $H$ satisfies conditions similar to those which $G$

*Cf. M. Hukuhara [36].

satisfies. In particular, $H(t, 0, 0, \ldots, 0) = 0$. Furthermore, *$H$ is independent of $z^{(n)}$.*

In this case, equation (6.2) becomes

$$q_n(t)z^{(n)} = h_0(t) + H(t, z, z', \ldots, z^{(n-1)}), \tag{7.1}$$

and we can prove the following result: Let $\arg t = \theta$ be a given direction in the $t$-plane such that $a < \theta < b$. Then, there exists a subsector $\hat{\mathbb{S}}$ of $\mathbb{S}$ given by (5.2) such that $a < \hat{a} < \theta < \hat{b} < b$, and that equation (7.1) admits a solution in the space $B(\hat{\mathbb{S}})$. The positive number $R$ must be sufficiently large.

Set

$$H(t, z, z', \ldots, z^{(n-1)}) = \sum_{k=0}^{n-1} q_k(t)z^{(k)} + \hat{H}(t, z, z', \ldots, z^{(n-1)}),$$

where $\hat{H}$ denotes those terms which are non-linear in $z$, $z', \ldots, z^{(n-1)}$. The linear part of equation (7.1) is given by

$$L(z) = q_n(t)z^{(n)} - \sum_{k=0}^{n-1} q_k(t)z^{(k)},$$

and differential equation (7.1) can be written as

$$L(z) = h_0(t) + \hat{H}(t, z, z', \ldots, z^{(n-1)}). \tag{7.2}$$

By using solutions of the linear homogeneous differential equation

$$L(z) = 0, \tag{7.3}$$

we reduce differential equation (7.2) to an integral equation so that a method with successive approximations can be utilized. To do this, we must fix a subsector $\hat{\mathbb{S}}$ of $\mathbb{S}$ in such a way that solutions of (7.3) *behave well* as $t$ tends to infinity in $\hat{\mathbb{S}}$. The structure of the general solution of (7.3) is well known*, and the

*Cf. W. Wasow [26; Chapters II, IV, and V], and E. A. Coddington and N. Levinson [8; Chapters 4 and 5]. As to irregular singularities, G. D. Birkhoff [28 and 29], and W. J. Trjitzinsky [51] are very important. See also M. Hukuhara [35, 36], J. Malmquist [42], and H. L. Turrittin [52].

sector $\hat{\mathfrak{S}}$ is determined by the behavior of solutions of (7.3) as $t$ tends to infinity in the given direction $\arg t = \theta$.*

## IV. ASYMPTOTIC EXPANSIONS OF A GIVEN SOLUTION

**8. Preliminaries.** Let us consider Problem I (cf. Section 2). Assume that $y = f(t)$ is a solution of differential equation (3.5). We want to find an asymptotic expansion of $f(t)$ as $t$ tends to infinity. In particular, we shall be concerned with the following questions:

(A) Does $f$ admit such an expansion? If $f$ admits such an expansion, what does it look like, roughly speaking?

(B) How can we find such an expansion exactly?

(C) Find such an expansion for each direction in which $t$ tends to infinity.

Problem (A) is the most fundamental of these three problems. Problems (B) and (C) are technical problems, but they are quite difficult as such.

As to Problem (A), it is worth-while to know that *almost all* entire functions of $t$ take all complex values in every sectorial domain (however small it may be).** "Almost all" means that such entire functions form a residual set in the space of entire functions with the topology of uniform convergence on each compact set. However, the set of all functions satisfying reasonable differential equations is not large. Therefore, Problem (A) is still a reasonable problem. It is very important to find a general criterion for those functions that admit asymptotic expansions as $t$ tends to infinity.

As to Problems (B) and (C), it should be noticed that, if $A_{sp}(f) = p$, then $p$ must be a formal solution of (3.5). (Cf. (4.3).) However, since formal solutions of (3.5) are not necessarily unique, a straightforward computation of formal solutions would not solve these problems.

---

*Cf. W. Wasow [26; Chapter IX], and B. Malgrange [41; Theorem 7.1, p. 164]. For the case when $n = 1$, a book by M. Hukuhara, T. Kimura, and T. Matuda (cf. [11]) provides us with a thorough and systematic method of finding various asymptotic results (in Chapters II and III).

**Cf. S. Kierst and E. Szipilrajn [38].

Let $\mathfrak{S}$ be a connected unbounded set in the $t$-plane. We denote by $O(1; \mathfrak{S})$ the set of all functions which are bounded in $\mathfrak{S}$. The set $O(1; \mathfrak{S})$ is an algebra with the unit-element, over **C**. The set $A(\mathfrak{S})$ is a subalgebra of $O(1; \mathfrak{S})$. (Cf. Section 4.) Let $\phi$ be a function which is defined in $\mathfrak{S}$. We set $O(\phi; \mathfrak{S}) = \{\phi g; g \in O(1; \mathfrak{S})\}$. If $\phi \in O(1; \mathfrak{S})$, then the set $O(\phi; \mathfrak{S})$ is an ideal of $O(1; \mathfrak{S})$. In general, $O(\phi; \mathfrak{S})$ is a vector space over **C**. Furthermore, if $g \in O(\phi; \mathfrak{S})$ and $h \in O(\psi; \mathfrak{S})$, then $gh \in O(\phi\psi; \mathfrak{S})$.

We also denote by $o(1; \mathfrak{S})$ the set of all functions which tend to zero as $t$ tends to infinity in $\mathfrak{S}$. The set $o(1; \mathfrak{S})$ is an ideal of $O(1; \mathfrak{S})$. The sets $O(t^{-m}; \mathfrak{S})$ $(m = 1, 2, \ldots)$ are ideals of $o(1; \mathfrak{S})$. In case when $\mathfrak{S}$ is a sector given by (5.1), the algebra $B(\mathfrak{S})$ is contained in $\cap_{m=0}^{\infty} O(t^{-m}; \mathfrak{S})$. (Cf. Section 6.) For any function $\phi$ which is defined in $\mathfrak{S}$, we set $o(\phi; \mathfrak{S}) = \{\phi g; g \in o(1; \mathfrak{S})\}$.

If we use these notations, the famous Stirling's formula becomes

$$\left( \frac{t^{\frac{1}{2}-t} e^t}{\sqrt{2\pi}} \right) \Gamma(t) - 1 \in O(t^{-1}; \mathfrak{S}), \tag{8.1}$$

where $\mathfrak{S}$ is the half-line $0 < t < +\infty$, and

$$\Gamma(t) = \int_0^{+\infty} u^{t-1} e^{-u}\, du \qquad \text{(i.e., Gamma function).} \tag{8.2}$$

If $f \in A(\mathfrak{S})$, and

$$A_{\text{sp}}(f) = \sum_{m=0}^{\infty} a_m t^{-m}, \tag{8.3}$$

then

$$f - \sum_{m=0}^{N} a_m t^{-m} \in O(t^{-N-1}; \mathfrak{S}) \quad \text{for} \quad N = 0, 1, \ldots. \tag{8.4}$$

Conversely, (8.4) implies (8.3). More generally speaking, a condition such as

$$f - \sum_{m=0}^{N} a_m t^{-m} \in O(t^{-p_N}; \mathfrak{S}) \quad \text{for} \quad N = 0, 1, \ldots,$$

implies (8.3), if $p_1 < p_2 < \cdots$ and $\lim_{N\to\infty} p_N = +\infty$.

The following two problems are more specific than Problem (A).

(A-1) Given $f$ and $\mathfrak{S}$, find $\phi$ such that

$$f \in O(\phi; \mathfrak{S}). \tag{8.5}$$

(A-2) Given $f$ and $\mathfrak{S}$, find $\phi$ such that

$$\frac{f}{\phi} - 1 \in o(1; \mathfrak{S}). \tag{8.6}$$

The function $\phi$ must be a well-known function. An asymptotic result such as (8.6) implies (8.5). However, the converse is not always true. Even when we get a result such as (8.5), it is frequently probable that $f \in o(\phi; \mathfrak{S})$. Under this situation, it is impossible to derive (8.6).

**9. The saddle point method.** In Section 2, we derived an asymptotic expansion (i.e., (2.4)) of function (2.1) by utilizing a repeated integration by parts. When a function is given by an integral, this is one of the powerful techniques.* Another well-known method is the saddle point method. This method applies to a function of a form

$$f(t) = \int_a^b e^{tg(u)} h(u)\, du. \tag{9.1}$$

Let $\mathfrak{S}$ be the half-line $0 < t < +\infty$, and let $\mathfrak{D}$ be a simply connected domain in the complex $u$-plane. Assume that $g(u)$ and $h(u)$ are holomorphic in $\mathfrak{D}$, and that the path of integration is taken from $a$ to $b$ within $\mathfrak{D}$. (The points $a$ and $b$ may be on the

*Cf. L. Sirovich [23; Chapter 2, §2.1].

boundary of $\mathfrak{D}$.) Since $g(u)$ and $h(u)$ are holomorphic, and since $\mathfrak{D}$ is simply connected, the path can be deformed without changing the value of $f(t)$. Any one of those paths is called an admissible path. For each admissible path $l$, set $\rho(l) = \max_{u \in l} \mathrm{Re}(g(u))$, if this exists. Assume that $\rho(l)$ is attained at a point $u_l$ on the path $l$. Let $u'$ and $u''$ be two points on $l$ such that $u_l$ is on the subarc $\widehat{u'\,u''}$. (Cf. Fig. 9.1.) Write $f(t)$ in the following form:

$$f(t) = e^{tg(u_l)}\left\{\int_{u'}^{u''} + \int_a^{u'} + \int_{u''}^{b} e^{t[g(u)-g(u_l)]}h(u)\,du\right\}. \qquad (9.2)$$

Assume that

$$\mathrm{Re}\big[g(u) - g(u_l)\big] \leqslant -\varepsilon_0 < 0 \qquad (9.3)$$

on the two arcs $\widehat{a\,u'}$ and $\widehat{u''\,b}$, where $\varepsilon_0$ is a positive constant. Note that $\mathrm{Re}(g(u_l)) = \rho(l)$. Under this situation, we may prove that

$$\int_a^{u'} + \int_{u''}^{b} e^{t[g(u)-g(u_l)]}h(u)\,du \in O(e^{-\varepsilon_0 t}, \mathfrak{S}), \qquad (9.4)$$

if the function $h(u)$ behaves well. Hence, term (9.4) is quite small as $t$ tends to $+\infty$, and the behavior of $f(t)$ as $t$ tends to $+\infty$ is determined mostly by the term

$$e^{tg(u_l)}\int_{u'}^{u''} e^{t[g(u)-g(u_l)]}h(u)\,du.$$

Since $\mathrm{Re}[g(u) - g(u_l)] \leqslant 0$, we get

$$f \in O(e^{tg(u_l)}, \mathfrak{S}). \qquad (9.5)$$

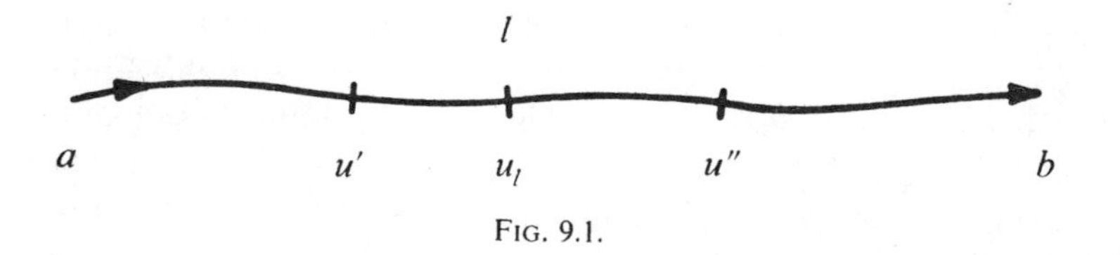


FIG. 9.1.

However, since this result depends on $l$, we might be able to improve this estimate by choosing some other paths. Set

$$\rho_0 = \min_{l} \max_{u \in l} \operatorname{Re}(g(u)), \tag{9.6}$$

where min is taken over all admissible paths. If this minimum $\rho_0$ is attained by an admissible path $l$, then formula (9.5) is the best we can get at this stage of the game. Formula (9.5) may be further improved through computation of an asymptotic expansion of the integral

$$\int_{u'}^{u''} e^{t[g(u)-g(u_l)]} h(u)\, du. \tag{9.7}$$

Therefore, this method consists of three steps:

(i) Find an optimal path which attains the optimal value $\rho_0$. (Cf. (9.6).)

(ii) Verify condition (9.3).

(iii) Compute an asymptotic expansion of (9.7) as $t$ tends to $+\infty$.

As to step (i), there are two possibilities:

($i$-$\alpha$) There is an admissible path $l$ such that $u_l = a$ (or $b$). In this case, $\rho_0 = \rho(l) = \operatorname{Re}(g(a))$ (or $\operatorname{Re}(g(b))$).

($i$-$\beta$) $u_l \neq a, b$ for every admissible path $l$.

In either case, it is worth-while to investigate the images of $\mathfrak{D}$ and $l$ by the mapping $w = g(u)$. Case ($i$-$\alpha$) is shown by Fig. 9.2. In case ($i$-$\beta$), there are many possibilities. If the situation is similar to Fig. 9.3, the optimal value $\rho_0$ might not be attained. However, instead of drawing such a quick conclusion, let us consider the case when $g(u) = u^2$ for an illustration of some other possibilities. Let $\xi$ and $\eta$ be the real and imaginary parts of $u$ (i.e., $u = \xi + i\eta$). Then $g(u) = (\xi^2 - \eta^2) + 2\xi\eta i$, and the two sectors defined by $\xi^2 - \eta^2 < 0$ in the $u$-plane are mapped onto the left half-plane in the $w$-plane. (Cf. Fig. 9.4.) Take two points $a$ and $b$ so that they are in the different sectors in the $u$-plane. Under this situation, the image of every admissible path must go around the point $w = 0$ in the $w$-plane. (Cf. Fig. 9.4.) Therefore, in this case, the optimal value $\rho_0$ is attained by a path which goes through the point $u = 0$.

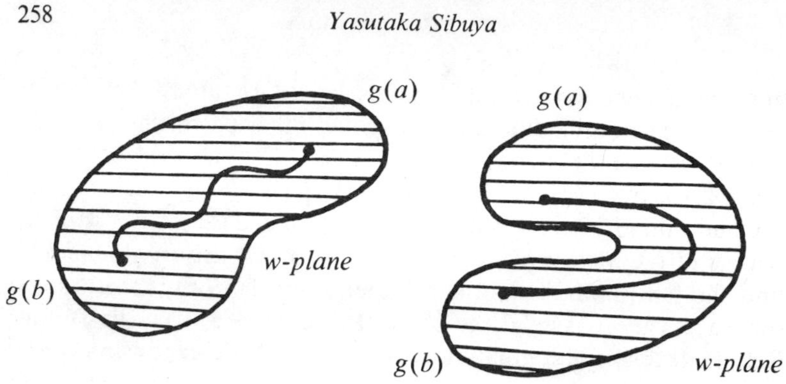


FIG. 9.2.

FIG. 9.3.

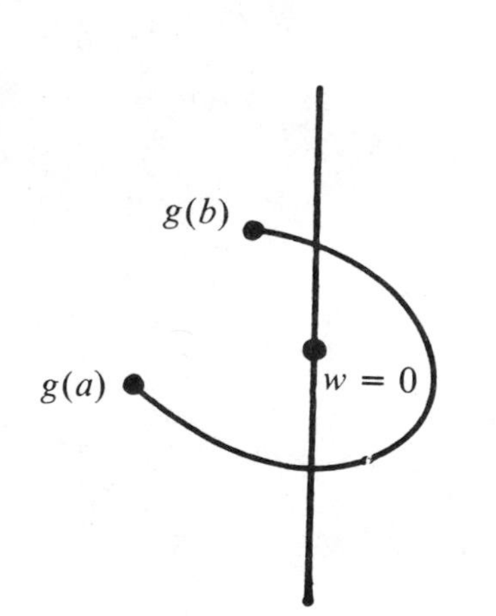


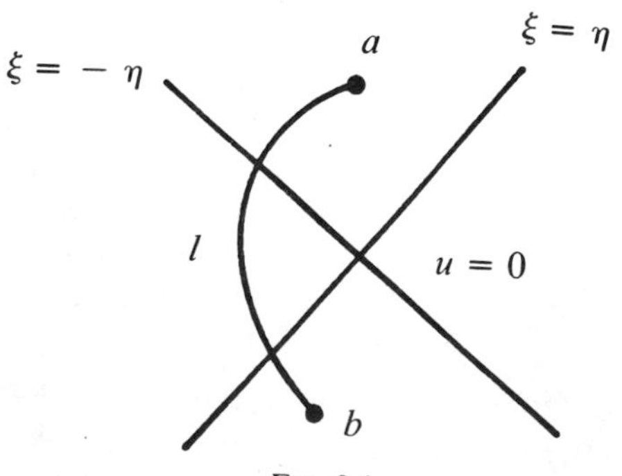


FIG. 9.4.

In general, since $\mathrm{Re}(g(u))$ is harmonic, the optimal value $\rho_0$ would be attained (most likely) by an admissible path which goes through one of zeros of $g'(u)$ (i.e., saddle points). Thus the first step is to locate zeros of $g'(u)$.

Condition (9.3) is a global condition on the path $l$. In general, it is very difficult to verify condition (9.3). A good suggestion is to find the family of curves defined by $\mathrm{Im}(g(u)) = $ constant. Along such a curve, it is relatively easy to investigate the behavior of $\mathrm{Re}(g(u))$. Note that curves defined by $\mathrm{Im}(g(u)) = $ constant are orbits of the differential equation

$$du/ds = \overline{g'(u)}, \tag{9.8}$$

where $\overline{\quad\quad}$ denotes the conjugate complex. If we take the real and imaginary parts of $u$ into account, differential equation (9.8) is an autonomous system in the $u$-plane.* Let $u_0$ be a zero of $g'(u)$. Suppose that a curve $l$ defined by $\mathrm{Im}(g(u)) = \mathrm{Im}(g(u_0))$ is an admissible path, and $u_l = u_0$. Suppose also that condition (9.3) is verified at least in a neighborhood of $u_0$, and that there are no zeros of $g'(u)$ on this curve $l$ other than $u_0$. Then, if we are in case ($i$-$\beta$), this curve satisfies requirements (i) and (ii). (The converse is not true.)

If we fix a path by (i) and (ii), the computation of an asymptotic expansion of (9.7) is rather a matter of routine. Usually, in such a computation, we choose the arc $\widehat{u'\,u''}$ so that $\mathrm{Im}(g(u)) =$ constant. Because of this, the saddle point method is also called the method of steepest descents.** In case when $\mathrm{Re}(g(u_l)) = \rho_0$, $g'(u_l) = 0$, and $g''(u_l) \neq 0$, function (9.1) admits the following formula:

$$\frac{e^{-tg(u_l)}}{h(u_l)} \left\{ \frac{-tg''(u_l)}{2\pi} \right\}^{\frac{1}{2}} f(t) - 1 \in O(t^{-1}, \mathfrak{S}),$$

*The study of curves $\mathrm{Im}(g(u)) = $ constant is very similar to that of orbits in the plane. See, for example, J. A. Jenkins [13; Chapter III], and M. A. Evgrafov and M. V. Fedoryuk [31].

**Cf. L. Sirovich [23; Chapter 2, §2.6], N. G. deBruijn [4; Chapter 5], and F. W. J. Olver [43].

where $\mathfrak{S}$ is the half-line $0 < t < +\infty$, and $-\arg((-tg''(u_l))^{\frac{1}{2}})$ must be the directional angle of the path at the saddle point $u_l$.

As an illustration, we consider Airy's equation

$$d^2y/dz^2 - zy = 0 \tag{9.9}$$

and its solution

$$Ai(z) = \frac{1}{2\pi i}\int_l e^{zv - \frac{1}{3}v^3}\,dv, \tag{9.10}$$

where a typical admissible path $l$ is shown by Fig. 9.5. Solution (9.10) is obtained through an application of Laplace transform. Set $\arg z = \theta$, $t = |z|^{\frac{3}{2}}$, and $u = v/\sqrt{|z|}$. Then

$$Ai(z) = \frac{\sqrt{|z|}}{2\pi i}\int_l e^{tg(u,\theta)}\,du,$$

where $g(u,\theta) = e^{i\theta}u - \frac{1}{3}u^3$, and $l$ is an admissible path given by Fig. 9.5. Observe that $g'(u,\theta) = e^{i\theta} - u^2$. Hence, $g'$ has two zeros $u_{\pm} = \pm e^{i\frac{1}{2}\theta}$. For $|\theta| < \pi$, one of the zeros (i.e, $u_-$) is in the left half-plane. There is an optimal admissible path which goes through the saddle point $u_-$ if $|\theta| < \pi$.

The curves $\mathrm{Im}(g(u,\theta)) =$ constant which go through the saddle points $u_+$ and $u_-$ are exhibited (symbolically) by Figs. 9.6, 9.7, 9.8, and 9.9 for various values of $\theta$. The arrow indicates the direction in which $\mathrm{Re}(g(u,\theta)$ increases. From these pictures, we may find an optimal admissible path $l$ which goes through $u_-$. In case $0 \leqslant \theta < 2\pi/3$, such a curve is given by $\mathrm{Im}(g(u,\theta)) = \mathrm{Im}(g(u_-,\theta))$. For $2\pi/3 \leqslant \theta < \pi$, we must use an arc $\mathrm{Re}(g(u,\theta)) =$ constant together with curves $\mathrm{Im}(g(u,\theta)) =$ constant. In case $\theta = \pi$, Fig. 9.9 still applies. However, in this case, $u_+$ and $u_-$ are joined by an arc $\mathrm{Re}(g(u,\theta)) =$ constant. Due to this situation, there is no optimal path going through $u_-$ if $\theta = \pi$. Note that these figures also exhibit the structural instability in case when two saddle points are connected by an orbit. At the end of all computations, we are able to derive a formula such as

$$F(z) - 1 \in O\left(z^{-\frac{3}{2}};\, \mathfrak{S}_\delta\right), \tag{9.11}$$

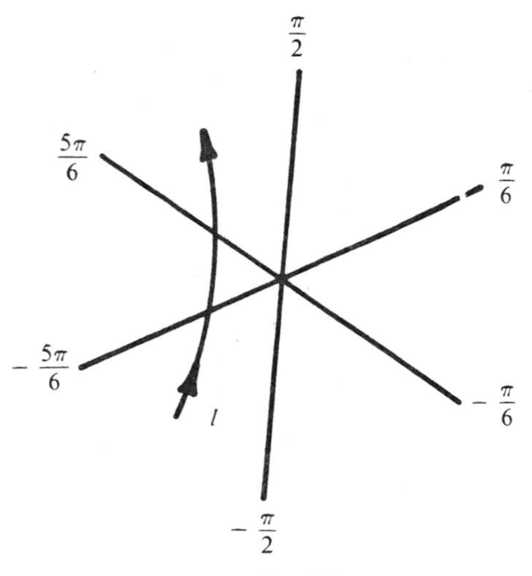


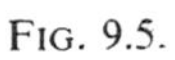
FIG. 9.5.

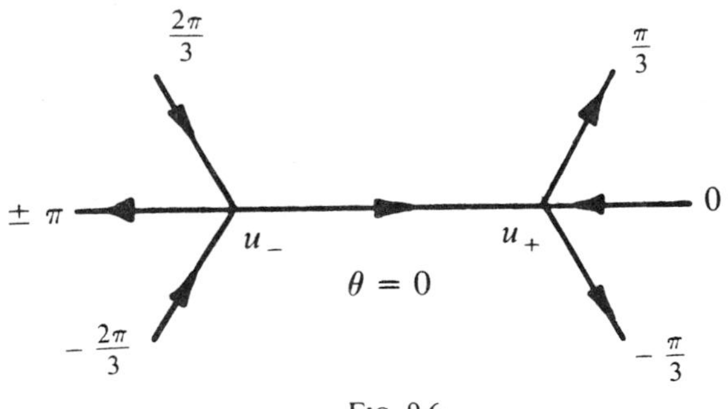


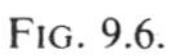
FIG. 9.6.

where

$$F(z) = \left( \frac{e^{-\frac{2}{3} z^{\frac{3}{2}}}}{2\sqrt{\pi}\, z^{\frac{1}{4}}} \right)^{-1} Ai(z), \tag{9.12}$$

and $\mathfrak{S}_\delta$ is a sector defined by $|\arg z| \leqslant \pi - \delta$ for an arbitrary

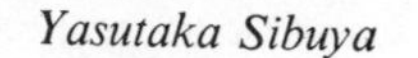

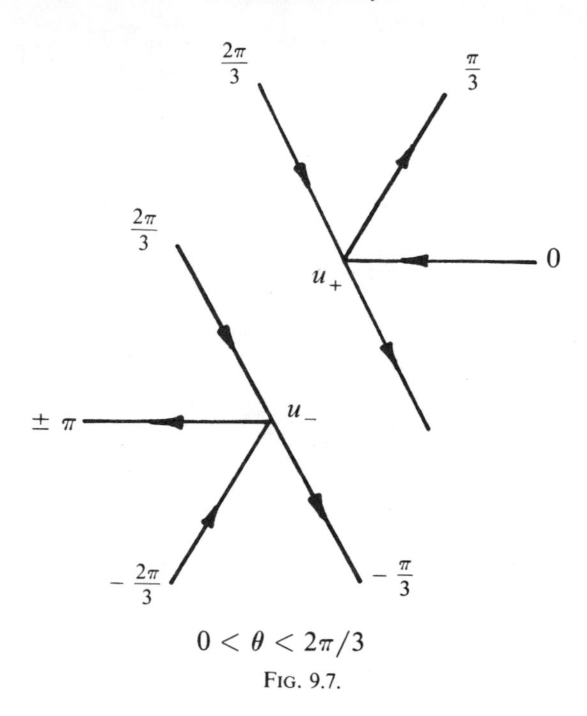


$0 < \theta < 2\pi/3$

FIG. 9.7.

small positive number $\delta$. We can also derive an asymptotic expansion of $F$ as $z$ tends to infinity in $\mathfrak{S}_\delta$.*

**10. Connection formulas.** The asymptotic expansion of function (9.12) can be computed also directly from differential equation (9.9) when once we derive (9.11). As we mentioned before, a straightforward computation of formal solutions would not solve Problems (B) and (C). (Cf. Section 8.) This is due to the fact that formal solutions depend on several arbitrary constants. Unless these constants are specified, formal solutions are not determined uniquely. An asymptotic result such as (9.11) fixes such constants. In fact, if we set $A_{\text{sp}}(F) = \Sigma_{m=0}^{\infty} a_m z^{-\frac{3}{2}m}$ with $a_0 = 1$, then the $a_m$ ($m \geqslant 2$) are uniquely determined by differential equation (9.9).

*For more applications, see, for example, F. W. J. Olver [19].

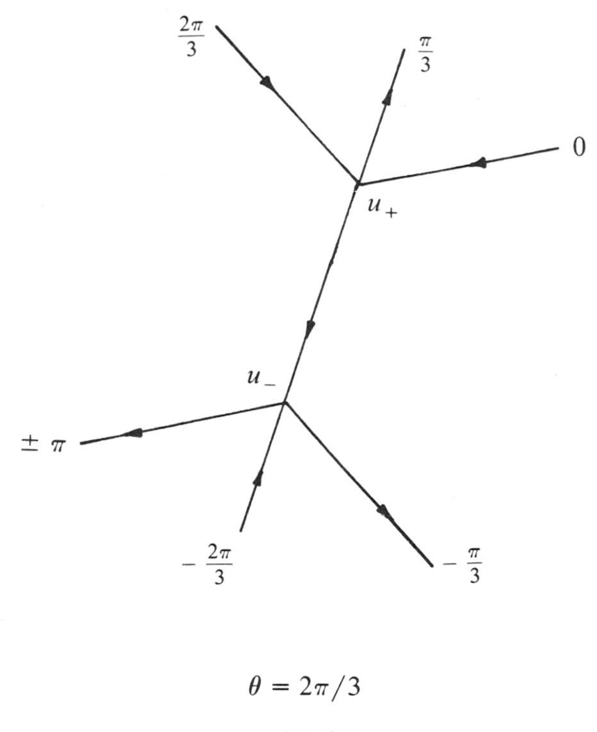


$$\theta = 2\pi/3$$

FIG. 9.8.

In general, for each formal solution, we may be able to construct a solution in such a way that the given formal solution is an asymptotic expansion of this solution. (Cf. Problem II, Section 2.) Under such a situation, if once we find a relation between a given solution and those solutions of known asymptotic behaviors, we can also find the asymptotic behavior of the given solution. Such a relation is called a connection formula.*

As the first illustration, we consider differential equation (2.2) (i.e., $y' - y = t^{-1}$). We know that function (2.1) is a solution, and that (2.4) is its asymptotic expansion as $t$ tends to $+\infty$. The

*As to those ideas, G. G. Stokes [48 and 49] are very important classical papers. See, also, R. Langer [40] and W. Wasow [53].

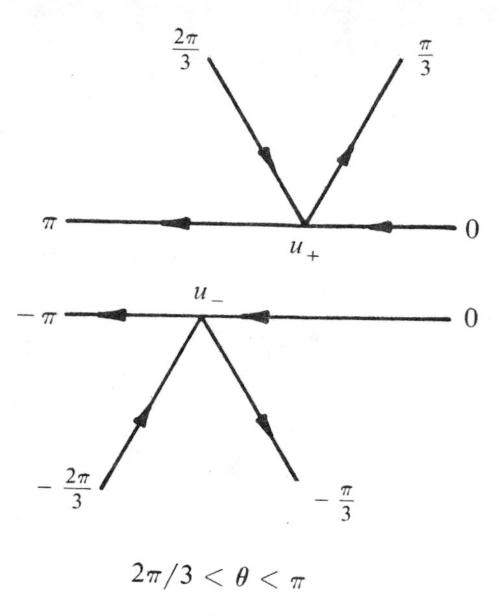


$2\pi/3 < \theta < \pi$

FIG. 9.9

function

$$g(t) = e^t \log t - \sum_{n=1}^{\infty} \left[ \sum_{k=1}^{n} \frac{1}{k} \right] \frac{t^n}{n!} \tag{10.1}$$

is also a solution of (2.2). Let us find the behavior of $g(t)$ as $t$ tends to $+\infty$. Observe that the general solution of (2.2) is given by $Ce^t + f(t)$ with an arbitrary constant $C$. Therefore, there must be a constant $C_0$ such that $g(t) = C_0 e^t + f(t)$. The main concern is to find $C_0$. Note that $A_{\text{sp}}(e^{-t}g) = C_0$ as $t$ tends to $+\infty$. If we set $w = e^{-t}y$, differential equation (2.2) becomes $w' = t^{-1}e^{-t}$. Since $A_{\text{sp}}(t^{-1}e^{-t}) = \mathbf{0}$, formal solutions of this equation are determined by $p' = \mathbf{0}$. This means that any constant $C \cdot \mathbf{1}$ is a formal solution. We want to find a particular formal solution $C_0 \cdot \mathbf{1}$. Instead of computing $C_0$ by $\lim_{t \to +\infty} e^{-t}g(t)$, let us observe that an integration by parts yields

$$f(t) = \log t - e^t \int_t^{+\infty} e^{-s} (\log s)\, ds,$$

and

$$C_0 e^t = g(t) - f(t) = (e^t - 1)\log t - \sum_{n=1}^{\infty} \left[ \sum_{k=1}^{n} \frac{1}{k} \right] \frac{t^n}{n!}$$

$$+ e^t \int_t^{+\infty} e^{-s} (\log s)\, ds. \tag{10.2}$$

By using formula (10.2) at $t = 0$, we obtain

$$C_0 = \int_0^{+\infty} e^{-s} (\log s)\, ds = -\gamma,$$

where $\gamma$ is Euler's constant (i.e., $\gamma = \lim_{n \to +\infty} \{\sum_{k=1}^{n} \frac{1}{k} - \log n\}$). Thus we get

$$g(t) = -\gamma e^t + f(t). \tag{10.3}$$

This is a connection formula. It follows from this formula that

$$A_{sp}(g + \gamma e^t) = \sum_{n=0}^{\infty} (-1)^{n+1} \frac{n!}{t^{n+1}} \tag{10.4}$$

as $t$ tends to $+\infty$.*

*The same idea may be used to derive an asymptotic expansion of $\sum_{m=0}^{\infty} t^m / \Gamma(m + \alpha)$. More generally, suppose that a function $a(m)$ satisfies a difference equation and $f(t) = \sum_{m=0}^{\infty} a(m) t^m$ satisfies a differential equation. Then, we may use either the differential equation or the difference equation in deriving an asymptotic expansion of $f(t)$. See, for example, W. B. Ford [10], B. L. J. Braaksma [30], M. Hukuhara [37], and W. Wasow [26; Chapter VI, §21].

In this computation, we utilized

(1) the general structure of solutions of differential equation (2.2),

(2) the behaviors of $f(t)$ and $g(t)$ in the neighborhood of $t = 0$,

(3) the behavior of $f(t)$ as $t$ tends to $+\infty$.

It was very important that the behaviors of $f(t)$ at $t = 0$ and $t$ tends to $+\infty$ were known. In general, when a solution is given by an integral, such an expression is helpful in deriving necessary formulas at various values of $t$. For example, Taylor expansion of $Ai(z)$ (cf. (9.10)) can be derived without any difficulty at $z = 0$. In particular

$$Ai(0) = \frac{1}{3^{\frac{2}{3}}\Gamma\left(\frac{2}{3}\right)}, \quad Ai'(0) = -\frac{1}{3^{\frac{1}{3}}\Gamma\left(\frac{1}{3}\right)}.$$

As the second illustration, let us consider Airy's equation (9.9). Observe that this equation is invariant by the change of the independent variable $z = \omega\zeta$, where $\omega = e^{i\frac{2}{3}\pi}$. This means that $Ai(\omega z)$ and $Ai(\omega^{-1}z)$ are solutions of (9.9). Note that $Ai(\omega z)$ and $Ai(\omega^{-1}z)$ satisfy, at $z = 0$, the initial conditions

$$y = Ai(0), \quad y' = \omega Ai'(0),$$

and

$$y = Ai(0), \quad y' = \omega^{-1}Ai'(0),$$

respectively. It follows that $Ai(\omega z)$ and $Ai(\omega^{-1}z)$ are linearly independent. Therefore, there must be two constants $c_1$ and $c_2$ such that $Ai(z) = c_1 Ai(\omega z) + c_2 Ai(\omega^{-1}z)$. Let $W_1$, $W_2$, and $W_3$ be the Wronskians of $\{Ai(z), Ai(\omega z)\}$, $\{Ai(z), Ai(\omega^{-1}z)\}$, and $\{Ai(\omega z), Ai(\omega^{-1}z)\}$, respectively. Then

$$c_1 = \frac{W_2}{W_3} = -\omega, \quad c_2 = -\frac{W_1}{W_3} = -\omega^{-1}.$$

Thus we obtained a connection formula

$$Ai(z) = -\omega Ai(\omega z) - \omega^{-1}Ai(\omega^{-1}z). \tag{10.5}$$

In this computation, we utilized

(4) an invariant property of differential equation (9.9),

(5) relations between two constants ($c_1$ and $c_2$) and the Wronskians of solutions involved.

Generally speaking, connection formulas are very difficult to find. Therefore, a wise suggestion is to find some way to simplify computations. For this purpose, invariant properties of differential equations under certain transformations are very useful. On the other hand, as far as linear homogeneous differential equations are concerned, computations of constants such as $c_1$ and $c_2$ are essentially those of ratios of Wronskians. Such a ratio is independent of $t$. Therefore, we may choose a value of $t$ so that the computation becomes as simple as we want. (The simplest way of deriving (10.5) is, of course, to use a particular form of integral (9.10).)

Note that the sector $\mathfrak{S}_\delta$ does not contain the negative real axis of the $z$-plane. This means that no information of the behavior of $Ai(z)$ as $z$ tends to $-\infty$ may be obtained directly from (9.11). For such an information, we can use formula (10.5). In fact, from formula (9.11), we derive the following formulas:

$$F(\omega z) - 1 \in O\left(z^{-\frac{3}{2}};\, \omega^{-1}\mathfrak{S}_\delta\right),$$

and

$$F(\omega^{-1} z) - 1 \in O\left(z^{-\frac{3}{2}};\, \omega\mathfrak{S}_\delta\right),$$

where $\omega^{\pm 1}\mathfrak{S}_\delta = \{\omega^{\pm 1}z;\, z \in \mathfrak{S}_\delta\}$. Note that the sectors $\omega^{\pm 1}\mathfrak{S}_\delta$ do contain the directions $\arg z = \pi$ and $\arg z = -\pi$ respectively. Therefore, from (10.5), we can derive information of the behavior of $Ai(z)$ as $z$ tends to $-\infty$. Since formula (9.11) involves many-valued functions such as $z^{-\frac{3}{2}}$, correct branches of those functions must be used in computations with formula (10.5).

Stirling's formula (8.1) is valid as $t$ tends to infinity in any sector $\mathfrak{S}_\delta$ (i.e., $|\arg t| \leqslant \pi - \delta$). However, no information of the behaviour of $\Gamma(t)$ (or $1/\Gamma(t)$) as $t$ tends to $-\infty$ may be derived directly from (8.1). For such information, another well-known formula $\Gamma(t)\Gamma(1-t)=\pi/\sin(\pi t)$, or

$$\frac{\pi}{\Gamma(t)}=\Gamma(1-t)\sin(\pi t) \tag{10.6}$$

is used. This is a connection formula concerning the difference equation $y(t+1)=ty(t)$.

**11. Simplification.** As we mentioned in Section 10 (cf. (1)), any information concerning the general structure of solutions of the given differential equation is useful in deriving an asymptotic expansion of a given solution. Such information may be obtained by constructing formal solutions systematically and by solving Problem II (cf. Section 2) for all formal solutions. For this purpose, certain preliminary observations are also helpful. Such observations are frequently based on the structure of solutions of well-known differential equations. For example, let us consider a differential equation

$$d^2y/dt^2-\rho y=q(t)y, \tag{11.1}$$

where $t$ is a real (or complex) independent variable, $\rho$ is a positive constant, and $q(t)$ is a function of $t$ which is defined for $t>0$. To solve this equation, we regard (11.1) as a perturbation of the equation

$$d^2y/dt^2-\rho y=0, \tag{11.2}$$

and we reduce equation (11.1) to an integral equation

$$y(t)=Ae^{\lambda t}+Be^{-\lambda t}$$
$$+\frac{1}{2\lambda}\int_{t_0}^{t}\{e^{\lambda(t-s)}-e^{-\lambda(t-s)}\}q(s)y(s)\,ds, \tag{11.3}$$

where $\lambda=\sqrt{\rho}$, and $A$ and $B$ are arbitrary constants. If $q(t)$ behaves well as $t$ tends to $+\infty$, we may be able to derive two independent solutions $y_1(t)$ and $y_2(t)$ such that

$$\begin{cases} e^{-\lambda t}y_1(t) - 1 \in o(1; \mathfrak{S}), \\ e^{\lambda t}y_2(t) - 1 \in o(1; \mathfrak{S}), \end{cases}$$

where $\mathfrak{S}$ is the half-line $0 < t < +\infty$. For a more general differential equation

$$d^2u/d\tau^2 - \rho p(\tau)u = 0, \tag{11.4}$$

where $\tau$ is an independent variable, and $p(\tau)$ is a function of $\tau$ for $\tau > 0$, we may use the transformation

$$t = \int_{\tau_0}^{\tau} \sqrt{p(s)}\, ds, \quad y = \sqrt[4]{p(\tau)}\, u \tag{11.5}$$

to derive an equation (11.1) from (11.4), where

$$q(t) = \frac{1}{4p(\tau)^2}\, p''(\tau) - \frac{5}{16p(\tau)^3}\, p'(\tau)^2. \tag{11.6}$$

Therefore, if the variables $t$ and $y$ are properly defined through (11.5), and if $q(t)$ behaves well as $t$ tends to $+\infty$, we may be able to construct two linearly independent solutions $u_1(\tau)$ and $u_2(\tau)$ of (11.4) such that

$$\begin{cases} \dfrac{\sqrt[4]{p(\tau)}\, u_1(\tau)}{\exp\left\{\lambda \displaystyle\int_{\tau_0}^{\tau} \sqrt{p(s)}\, ds\right\}} - 1 \in o(1; \mathfrak{S}), \\ \dfrac{\sqrt[4]{p(\tau)}\, u_2(\tau)}{\exp\left\{-\lambda \displaystyle\int_{\tau_0}^{\tau} \sqrt{p(s)}\, ds\right\}} - 1 \in o(1; \mathfrak{S}). \end{cases}$$

In this computation, we utilized the structure of solutions of (11.2) and the validity of transformation (11.5). In other words, we simplified equation (11.4) to (11.1) by (11.5), and we utilized the structure of solutions of (11.2). In case when $p(\tau)$ vanishes at a point (or points) in the interval under consideration, transformation (11.5) may not be used properly. In such a case, differential equation (11.2) would not provide us with satisfactory information. We must replace (11.2) with a more general differential equation.* For example, Airy's equation (9.9) is one of those useful equations. Various special functions are also useful.** The lack of knowledge concerning differential equations of simple forms, however, is still one of the serious obstacles for the applications of asymptotic methods. Therefore, it is worthwhile to collect such information.***

For simplification of a given differential equation, there are some guide-lines. A typical result will be given below. Let us consider a system of linear differential equations

$$dw/dt = \left[\Lambda(t) + U(t)\right]w, \tag{11.7}$$

where $w$ is an $n$-dimensional vector, and $\Lambda(t) + U(t)$ is an $n$-by-$n$ matrix. Assume that $\Lambda(t)$ is a diagonal matrix with diagonal elements $\lambda_1(t), \ldots, \lambda_n(t)$ satisfying the conditions:
(a) $\lambda_1(t), \ldots, \lambda_n(t)$ are continuous for $0 \leqslant t < +\infty$,
(b) $\mathrm{Re}(\lambda_{j+1}(t)) \geqslant \mathrm{Re}(\lambda_j(t))$ for $t \geqslant 0, j = 1, \ldots, n-1$,
(c) $\lim_{t\to+\infty}\int_0^t \mathrm{Re}[\lambda_{j+1}(s) - \lambda_j(s)]\,ds = +\infty$ $(j = 1, \ldots, n-1)$.
Assume also that the elements of $U(t)$ are continuous for $t \geqslant 0$, and that

$$\int_0^{+\infty} |U(t)|\,dt < +\infty. \tag{11.8}$$

---

*This is a starting point of the problem of transition points (or turning points). The concept of "related equations" was thus introduced by R. E. Langer [39]. For transition points, see A. Erdélyi [9], and W. Wasow [26; Chapter VIII].

**Cf. F. W. J. Olver [19].

***See, for example, C. A. Swanson and V. B. Headley [50], P. F. Hsieh and Y. Sibuya [34], and Y. Sibuya [47].

Then, there exists an $n$-by-$n$ matrix $P(t)$ such that $P(t)$ tends to the identity matrix as $t$ tends to $+\infty$, and that the transformation $w = P(t)u$ reduces system (11.7) to $du/dt = \Lambda(t)u$. There are many variations of this result.*

**12. Remarks.** The growth properties of meromorphic functions as $t$ tends to infinity are closely related with the distributions of their values. A meromorphic function $f(t)$ has a tendency that, if $f(t)$ cannot take a particular value $c$ so many times as expected, then $f(t)$ takes $c$ at $t = \infty$ so that the deficiency is made up for. As an example, let us consider a polynomial $p(t)$ in $t$ of degree $n$. This function takes every finite complex value $n$ times. However, this function cannot take $\infty$ at any finite value of $t$. *Hence*, $\infty$ is taken $n$ times at $t = \infty$. This *explains* a reason why $p(t) \in O(t^n)$ as $t$ tends to infinity. As another example, let us consider $e^t$. This function does not take 0 and $\infty$ at any finite value of $t$, although other values are taken infinitely many times. This *explains* a reason why $e^t$ tends to 0 and $\infty$ as $t$ tends to infinity. This relation between the growth property and the distribution of values of a meromorphic function may be beautifully explained by the theory of R. Nevanlinna.**

For an entire function $f(t)$ in $t$, its order $\rho(f)$ is defined by

$$\rho(f) = \overline{\lim_{r \to +\infty}} \frac{\log \log M(r)}{\log r},$$

where

$$M(r) = \max_{|t| \leqslant r} |f(t)|.$$

If $\rho(f) < \rho(g)$, then $g$ takes almost all finite values more often than $f$ does. Another interesting fact is that, if $\rho(f)$ is finite and if

*Cf. E. A. Coddington and N. Levinson [8; Chapter 3, §8, Theorem 8.1, p. 92], L. Césari [7; Chapter II], R. Bellman [1; Chapter 2], W. A. Harris and D. A. Lutz [33].

**Cf. R. Nevanlinna [18]. For applications of Nevanlinna's theory to ordinary differential equations, see H. Wittich [27; Chapter V].

$\rho(f)$ is not an integer, $f$ takes every finite value infinitely many times. Assume that $\rho(f) = \rho_0 < +\infty$, and that $f(t)$ tends to $c_1(\neq \infty)$ and $c_2(\neq \infty)$ as $t$ tends to infinity in two directions $\arg t = \theta_1$ and $\arg t = \theta_2$, respectively. Then, if $0 < \theta_1 - \theta_2 < \pi/\rho_0$, we have $c_1 = c_2$. Furthermore, $f(t)$ tends to $c_1(= c_2)$ uniformly as $t$ tends to infinity in the sector $\theta_2 \leqslant \arg t \leqslant \theta_1$. In general, an entire function with a regular pattern of distribution of values admits an asymptotic representation as $t$ tends to infinity.* Conversely, an entire function with a reasonable asymptotic property usually exhibits a regular pattern of distribution of values.

In this article were considered asymptotic properties as $t$ tends to infinity. Similar properties may be treated in the same manner as $t$ tends to a finite value. Also, asymptotic properties with respect to a parameter may be studied in the same manner. The problem of singular perturbations is one of the most significant applications of asymptotic methods with respect to an independent variable (i.e., $t$) as well as a parameter. In particular, "boundary layer terms" make the problem more interesting and difficult.** Another big area of applications is the study of spectra of linear differential operators.*** In many applications, asymptotic methods may be utilized to prove the existence of solutions satisfying certain asymptotic conditions. For this purpose, various methods for Problems I and II (cf. Section 2) must be refined and combined together. Also, asymptotic properties with respect to an independent variable and a parameter (or parameters) must be considered simultaneously.

According to Plato, the material world is a shadow of the world of ideals on the walls of the cave in which we are imprisoned in this life. The world of ideals may be conceived only asymptotically. The material world is imperfect. It was written somewhere that the great pyramid of Giza was built very accurately. But, the longest side differs from the shortest by 7.9 inches. "Asymptotic methods" will be indispensable for ever. Through justification of

---

*Cf. B. Ja. Levin [15; Chapter I, II, and III]. For entire functions, see also R. P. Boas [2].

**Cf. W. Wasow [26; Chapter X], and R. E. O'Malley [20].

***Cf. E. C. Titchmarsh [24], and M. A. Neumark [17; Chapters II and VII].

asymptotic results, we will be able to understand how imperfect the material world is.*

## REFERENCES


(A) **Books:**

1. R. Bellman, *Stability Theory of Differential Equations*, McGraw-Hill, New York, 1953.
2. R. P. Boas, Jr., *Entire Functions*, Academic Press, New York, 1954.
3. É. Borel, *Leçons sur les séries divergentes*, 2nd ed., Gauthier-Villars, Paris, 1928.
4. N. G. de Bruijn, *Asymptotic Methods in Analysis*, 2nd ed., North-Holland, Amsterdam, 1961.
5. T. Carleman, *Leçons sur les fonctions quasi analytiques*, Gauthier-Villars, Paris, 1926.
6. H. Cartan, *Elementary Theory of Analytic Functions of One or Several Complex Variables*, Addison-Wesley, Reading, 1963.
7. L. Césari, *Asymptotic Behavior and Stability Problems in Ordinary Differential Equations*, Springer-Verlag, New York, 1959.
8. E. A. Coddington and N. Levinson, *Theory of Ordinary Differential Equations*, McGraw-Hill, New York, 1955.
9. A. Erdélyi, *Asymptotic Expansions*, Dover, New York, 1956.
10. W. B. Ford, *The Asymptotic Developments of Functions Defined by Maclaurin Series*, University of Michigan, 1936, (Chelsea, New York, 1960).
11. M. Hukuhara, T. Kimura, and T. Matuda, *Équations différentielles ordinaires du premier ordre dans le champ complexe*, The Mathematical Society of Japan, 1961.
12. H. Jeffreys, *Asymptotic Approximations*, Oxford University Press, New York, 1962.
13. J. A. Jenkins, *Univalent Functions and Conformal Mapping*, Springer-Verlag, New York, 1965, (2nd printing corrected).
14. E. R. Kolchin, *Differential Algebra and Algebraic Groups*, Academic Press, New York, 1973.
15. B. Ja. Levin, *Distribution of Zeros of Entire Functions*, American Mathematical Society, 1964.
16. A. H. Nayfeh, *Perturbation Methods*, Wiley, New York, 1973.


---

*For applications of asymptotic methods, see K. O. Friedrichs [32].

17. M. A. Neumark, *Linear Differential Operators*, Parts I and II, Frederick Ungar, New York, 1968.

18. R. Nevanlinna, *Analytic Functions*, Springer-Verlag, New York, 1970.

19. F. W. J. Olver, *Asymptotics and Special Functions*, Academic Press, New York, 1973.

20. R. E. O'Malley, Jr., *Introduction to Singular Perturbations*, Academic Press, New York, 1974.

21. F. Pittnauer, *Vorlesungen über asymptotische Reihen*, Lecture Notes in Mathematics 301, Springer-Verlag, New York, 1972.

22. H. Poincaré, *Les méthodes nouvelles de la mécanique céleste*, Gauthier-Villars, Paris, 1893.

23. L. Sirovich, *Techniques of Asymptotic Analysis*, Springer-Verlag, New York, 1971.

24. E. C. Titchmarsh, *Eigenfunction Expansions Associated with Second-Order Differential Equations*, Part One, 2nd ed., Oxford University Press, New York, 1962.

25. R. J. Walker, *Algebraic Curves*, Princeton University Press, Princeton, 1950.

26. W. Wasow, *Asymptotic Expansions for Ordinary Differential Equations*, Wiley, New York, 1965.

27. H. Wittich, *Neuere Untersuchungen über eindeutige analytische Funktionen*, Springer-Verlag, New York, 1955.

(B) **Papers:**

28. G. D. Birkhoff, "Singular points of ordinary linear differential equations," *Trans. Amer. Math. Soc.*, **10** (1909), 436–470.

29. ———, "The generalized Riemann problem for linear differential equations and the allied problems for linear difference and $q$-difference equations," *Proc. Amer. Acad. Arts and Sci.*, **49** (1913), 521–568.

30. B. L. J. Braaksma, "Asymptotic analysis of a differential equation of Turrittin," *SIAM J. Math. Anal.*, **2** (1971), 1–16.

31. M. A. Evgrafov and M. V. Fedoryuk, "Asymptotic behaviour as $\lambda \to \infty$ of the solution of the equation $w''(z) - p(z, \lambda)w(z) = 0$ in the complex $z$-plane," *Russian Math. Surveys*, **21** (1966), 1–48.

32. K. O. Friedrichs, "Asymptotic phenomena in mathematical physics," *Bull. Amer. Math. Soc.*, **61** (1955), 485–504.

33. W. A. Harris, Jr. and D. A. Lutz, "Recent results in the asymptotic integration of linear differential systems," *International Conference on differential equations*, Academic Press, New York, 1975, 345–349.

34. P. F. Hsieh and Y. Sibuya, "On the asymptotic integration of second order linear ordinary differential equations with polynomial coefficients," *J. Math. Anal. Appl.*, **16** (1966), 84–103.

35. M. Hukuhara, "Sur les points singuliers des équations différentielles linéaires, II," *J. Fac. Sci. Hokkaido Univ.*, **5** (1937), 157–166.

36. ———, "Sur les points singuliers des équations différentielles linéaires, III," *Mem. Fac. Sci. Kyushu Univ.*, **2** (1942), 125–137.

37. ———, "Sur le problème de connexion pour l'équation différentielle linéaire," *Boll. Un. Mat. Ital.*, **11**—Suppl. (1975), 87–94.

38. S. Kierst and E. Szipilrajn, "Sur certaines singularités des fonctions analytiques uniformes," *Fund. Math.*, **21** (1933), 276–294.

39. R. E. Langer, "The asymptotic solutions of ordinary linear differential equations of the second order, with special reference to a turning point," *Trans. Amer. Math. Soc.*, **67** (1949), 461–490.

40. ———, "The asymptotic solutions of ordinary linear differential equations of the second order, with special reference to the Stokes' phenomenon," *Bull. Amer. Math. Soc.*, **40** (1934), 545–582.

41. B. Malgrange, "Sur les points singuliers des équations différentielles," L'Enseignement Math., **20** (1974), 147–176.

42. J. Malmquist, "Sur l'étude analytique des solutions d'un système des équations différentielles dans le voisinage d'un point singulier d'indétermination, I, II, III," *Acta Math.*, **73** (1940), 87–129; **74** (1941), 1–64; **74** (1941), 109–128.

43. F. W. J. Olver, "Why steepest descents," *SIAM Rev.*, **12** (1970), 228–247.

44. H. Poincaré, "Sur les équations linéaires aux différentielles ordinaires et aux différences finies," *Amer. J. Math.*, **7** (1885), 1–56.

45. ———, "Sur les intégrales irrégulières des équations linéaires," *Acta Math.*, **8** (1886), 295–344.

46. ———, "Remarques sur les intégrales irrégulières des équations linéaires (réponse à M. Thomé)," *Acta Math.*, **10** (1887), 310–312.

47. Y. Sibuya, "Uniform simplification in a full neighborhood of a transition point," *Mem. Amer. Math. Soc.*, **149** (1974).

48. G. G. Stokes, "On the discontinuity of arbitrary constants which appear in divergent developments," *Trans. Cambridge Philos. Soc.*, **10** (1857), 106–128.

49. ———, "On the numerical calculation of a class of definite integrals and infinite series," *Trans. Cambridge Philos. Soc.*, **9** (1850), 163–191.

50. C. A. Swanson and V. B. Headley, "An extension of Airy's equation," *SIAM J. Appl. Math.*, **15** (1967), 1400–1412.

51. W. J. Trjitzinsky, "Analytic theory of linear differential equations," *Acta Math.*, **62** (1934), 167–226.

52. H. L. Turrittin, "Convergent solutions of ordinary linear homogeneous differential equations in the neighborhood of an irregular singular point," *Acta Math.*, **93** (1955), 27–66.

53. W. Wasow, "Connection problems for asymptotic series," *Bull. Amer. Math. Soc.*, **74** (1968), 831–853.

# INDEX